A NORTH CAROLINA GUIDE TO

Animal Control Law

AIMEE N. WALL

2008

THE UNIVERSITY
of NORTH CAROLINA
at CHAPEL HILL

The School of Government at the University of North Carolina at Chapel Hill works to improve the lives of North Carolinians by engaging in practical scholarship that helps public officials and citizens understand and strengthen state and local government. Established in 1931 as the Institute of Government, the School provides educational, advisory, and research services for state and local governments. The School of Government is also home to a nationally ranked graduate program in public administration and specialized centers focused on information technology, environmental finance, and civic education for youth.

As the largest university-based local government training, advisory, and research organization in the United States, the School of Government offers up to 200 classes, seminars, schools, and specialized conferences for more than 12,000 public officials each year. In addition, faculty members annually publish approximately fifty books, periodicals, and other reference works related to state and local government. Each day that the General Assembly is in session, the School produces the *Daily Bulletin*, which reports on the day's activities for members of the legislature and others who need to follow the course of legislation.

The Master of Public Administration Program is a full-time, two-year program that serves up to sixty students annually. It consistently ranks among the best public administration graduate programs in the country, particularly in city management. With courses ranging from public policy analysis to ethics and management, the program educates leaders for local, state, and federal governments and nonprofit organizations.

Operating support for the School of Government's programs and activities comes from many sources, including state appropriations, local government membership dues, private contributions, publication sales, course fees, and service contracts. Visit www.sog.unc.edu or call 919.966.5381 for more information on the School's courses, publications, programs, and services.

The School of Government does not discriminate in offering access to its educational programs and activities on the basis of age, gender, race, color, national origin, religion, creed, disability, veteran's status, or sexual orientation.

Printed in the United States of America
12 11 10 09 08 1 2 3 4 5
ISBN 978-1-56011-577-9
∞ This publication is printed on permanent, acid-free paper in compliance with the North Carolina General Statutes.
♲ Printed on recycled paper
Photographs courtesy of members of the faculty and staff of the School of Government.

Contents

Preface

In 1978 the Institute of Government published *The North Carolina Dog Manual.* The author, Patrice Solberg, introduced the topic by explaining that "no problem plagues local government officials more than the issues surrounding dog control." In later years, faculty members L. Poindexter Watts and Ben Loeb Jr. revised and expanded Ms. Solberg's work to address, under various titles, the full range of animal control issues confronting North Carolina's local governments.

The present publication, *A North Carolina Guide to Animal Control Law,* builds on this earlier work. It adds new material in the core areas of the law, including animal cruelty, dangerous dogs, and nuisance animals. It also introduces several entirely new areas of the law, including the regulation of animal shelters, the state's spay/neuter program, and laws governing assistance animals. The guide is designed primarily for those who work in the animal control field at the local level and for the attorneys who advise them. Pet owners and animal welfare organizations as well members of city and county governing boards and health directors may also find it useful.

Included at the end of each chapter are texts of selected relevant laws. It is important to remember that the laws in this field change often. For example, as this guide was being finalized, controversial state regulations governing euthanasia procedures in animal shelters were under review, and legislation that would significantly change the state's laws governing dangerous dogs was pending in the General Assembly. Officials charged with enforcing animal control laws therefore need to stay up-to-date with changes in those laws. To help them do so, supplements to this guide will be published periodically and timely updates will be available online at www.ncanimalcontrol.unc.edu.

I would like to extend sincere thanks to my most recent predecessor in this field, Ben Loeb. Before his retirement, Professor Loeb spent hours sharing stories, explaining the law, and providing much-needed guidance. I would also like to thank Harvey Barbee, who worked tirelessly as a research assistant to help fill in the gaps of my knowledge. Thanks are also due to Alex Hess and Marsha Lobacz of the Joseph P. Knapp Library at the School of Government. Alex and Marsha uncovered legislative history, tracked statutory developments, and identified other important historical trends in this field. I would also like to thank

Roberta Clark, Dan Soileau, Kevin Justice, and the rest of the Publications team for editing, designing, and organizing this new publication. Finally, I am most grateful to my husband, Steve Wall, who not only provided moral support but proofread every chapter of the manuscript.

Aimee N. Wall
Chapel Hill
June 2008

Chapter 1

Criminal Cruelty

A variety of legal tools are available at the federal, state, and local levels to address the abuse, neglect, or otherwise cruel treatment of animals. Under federal law, a few specific activities, such as animal fighting, are subject to criminal penalties. At the state level, North Carolina law includes both criminal and civil remedies for animal cruelty. The criminal laws target general acts of cruelty as well as such specific activities as animal fighting.[1] The civil laws allow any person to ask a court to enjoin another person from cruelly treating an animal.[2] Many cities and counties have also adopted ordinances that supplement the remedies available under federal and state law.[3]

This chapter—the first of two addressing animal cruelty laws—reviews the state's criminal cruelty laws in detail and discusses some of the court decisions that have shaped this area of the law. It also briefly discusses federal laws on animal cruelty. Chapter 2 addresses the state's civil remedies, local governments' authority to appoint animal cruelty investigators, and the laws governing those investigators. Neither chapter deals with the law governing an animal owner's ability to sue another person for money damages related to the loss or injury of an animal.

The earliest state criminal statutes governing animal cruelty were passed in the late nineteenth century. Many of those enacted by the General Assembly as far back as 1881 remain in force today, although amended a bit over time. The cornerstone of the cruelty laws is the general law prohibiting cruelty found in sections 14-360 and 14-361 of the North Carolina General Statutes (hereinafter G.S.).

---

1. Article 47 of N.C. Gen. Stat. Chapter 14 (hereinafter G.S.).
2. G.S. 19A-1 through 19A-9.
3. The General Assembly granted cities and counties specific statutory authority to "define and prohibit the abuse of animals." *See* G.S. 153A-127 (counties); G.S. 160A-182 (cities).

General Cruelty

North Carolina law recognizes two different levels of cruelty; one is punishable as a misdemeanor and the other as a felony. The primary differences between the two levels are (1) the defendant's state of mind and (2) the severity of the harm caused to the animal. Both levels of cruelty apply to *animals*, defined broadly to include "every living vertebrate in the classes Amphibia, Reptilia, Aves, and Mammalia except human beings."[4]

Misdemeanor Cruelty

It is a Class 1 misdemeanor to intentionally do any of the following:

- overdrive an animal
- overload an animal
- wound an animal
- injure an animal
- torment an animal
- kill an animal
- deprive an animal of necessary sustenance

In addition, it is a Class A1 misdemeanor (which carries higher penalties) to maliciously kill an animal by intentionally depriving it of necessary sustenance.[5]

A person who causes or procures an act resulting in one of the seven types of cruelty identified above can also be found guilty of a misdemeanor.[6] For example, if Person A causes Person B to intentionally kill an animal, both Person A and Person B can be charged with misdemeanors.

To understand the parameters of misdemeanor cruelty, it is important to examine closely the key terms used in the law. According to the statute, an

---

4. G.S. 14-360(c). The categories of animals protected under the law have changed over time. In 1881 the law applied to "any useful beast, fowl, or animal" and "animal" was defined to include "every living creature." N.C. Code §§ 2482, 2490 (1883). The language related to useful beasts was later dropped, but the general category of "every living creature" remained in the law until 1998, when it was changed to "every living vertebrate except human beings." S.L. 1998-212, sec. 17.16(c). In 1999 the definition took its current form. S.L. 1999-209, sec. 8.

The same definition is also used in the context of civil animal cruelty cases. *See* G.S. 19A-1(1). One scholar suggests that the definition should be expanded because it omits certain classes of animals, such as Pisces (fish). *See* William A. Reppy Jr., "Citizen Standing to Enforce Anti-Cruelty Laws by Obtaining Injunctions: The North Carolina Experience," *Animal Law* 11 (2005): 39, 45–46 (advocating a return to the more expansive definition).

5. G.S. 14-360(a1). This provision was added to the statute in 2007. S.L. 2007-211.

6. G.S. 14-360(a). The term *procure* means "to contrive, bring about, effect, or cause." *Black's Law Dictionary*, 6th ed. (1990), 1208.

act is done *intentionally* if it is committed knowingly and without justifiable excuse.[7] The term *knowingly* is not defined in the statute but has been interpreted by the courts in the context of other criminal laws to mean that the person is aware or conscious of what he or she is doing.[8] Combined, the two terms describe the state of mind required for a finding of misdemeanor cruelty: the person (1) is aware or conscious of what he or she is doing and (2) does not have a justifiable excuse.[9]

Whether an act was done without a justifiable excuse is often the subject of litigation. Basically, the court must decide if the person had a legally defensible reason for causing or permitting an animal's pain, suffering, or death. Courts have long recognized self-defense or defense of others as justifiable excuses.[10] They are less willing to recognize defense of property as an adequate justification. For example, the court in *State v. Neal* rejected one person's argument that he was justified in killing his neighbor's chickens because they were eating his family's pea crop.[11]

While an argument based on defense of crops is unlikely to succeed, it is possible to argue defense of property if the personal property in question is another animal. Nonetheless, the courts have been careful to narrowly limit the availability of this justification to situations in which it is necessary for an animal owner to kill or injure an animal to protect his or her own animal(s).[12] For

---

7. G.S. 14-360(c). Note that the Class 1 misdemeanor cruelty charge does not require evidence that the person acted maliciously or with evil intent.

8. Jessica Smith, *North Carolina Crimes: A Guidebook on Elements of Crime*, 6th ed. (Chapel Hill: UNC School of Government, 2008), 3 ("A person acts (or fails to act 'knowingly' when the person is aware of what he or she is doing. A person has knowledge of a condition . . . when he or she has actual information concerning the condition. The fact that the defendant had a reasonable belief of a fact is not sufficient to establish that the defendant acted knowingly; to establish this mental state the person must actually have knowledge of the fact." (internal citations omitted)).

9. The misdemeanor statute uses the term *maliciously* only once, when defining as guilty of a Class A1 misdemeanor someone who "shall maliciously kill, or cause or procure to be killed, any animal by intentional deprivation of necessary sustenance." G.S. 14-360(a1).

10. *See* State v. Simmons, 36 N.C. App. 54, 244 S.E.2d 168 (1978).

11. 120 N.C. 613, 27 S.E. 81 (1897); *see also* State v. Butts, 92 N.C. 784, 787, 1885 WL 1606 (N.C.), at *2 ("It never was the law that a man might shoot and kill his neighbor's horses and cows for a trespass upon his crops.").

12. *See* Parrott v. Hartsfield, 20 N.C. 242, 244, 1838 WL 523 (N.C.), at *2 ("The law authorizes the act of killing a dog found on a man's premises in the act of attempting to destroy his sheep, calves, coneys [rabbits] in a warren, deer in a park, or other reclaimed animals used for human food and unable to defend themselves. . . . The law is different where the dog is chasing animals *feræ naturæ,* such as hares or deer in a wild state, or combating with another dog."); State v. Dickens, 215 N.C. 303, 305, 1 S.E.2d 837, 839

example, in 2006 the Court of Appeals issued an unpublished opinion rejecting a defendant's argument that he was justified in killing a dog that was fighting with his dog. In this case, the defendant had stopped the fight but after doing so shot and killed the other dog.[13] The court recognized that (1) neither dog appeared to be the aggressor, (2) the dog killed had no history of aggression, (3) no animals or people were at risk after the fight was interrupted, and (4) the defendant was easily able to stop the fight. Taken together, the court held that these findings supported the state's argument that deadly force was not necessary to protect his dog or other people and therefore that the killing was not justified.

Numerous other cases dating back to the late nineteenth century address the issue of justification. The courts have found that the following are not legally sufficient justifications for acts of cruelty: [14]

- A person's "desire for amusement and sport"
- A person's "impulse of anger"
- An animal's previous offense, such as trespassing.

Although rather old, the cases cited interpret and apply statutes that are quite similar to North Carolina's current cruelty statute.[15] As such, they pro-

---

(1939) ("There was here no evidence offered that the dog of the prosecuting witness, at the time he was killed, was attempting to attack any animal or person, or threatening injury to property, so as to reasonably lead the defendant to believe that it was necessary to kill in order to protect the property of his employer."); *see also* State v. Smith, 156 N.C. 628, 634, 72 S.E. 321, 323 (1911) ("If the danger to the animal, whose injury or destruction is threatened, be imminent or his safety presently menaced, in the sense that a man of ordinary prudence would be reasonably led to believe that it is necessary for him to kill in order to protect his property, and to act at once, he may defend it, even unto the death of the dog, or other animal, which is about to attack it.").

13. State v. Dockery, 634 S.E.2d 641, 2006 WL 2671342 (N.C. App. Sept. 19, 2006) (unpublished decision).

14. State v. Porter, 112 N.C. 887, 16 S.E. 915 (1893) ("Since the enactment of [the cruelty] statute, it has been unlawful in this state for a man to gratify his angry passions or his love for amusement and sport at the cost of wounds and death to any useful creature over which he has control."); *Neal,* 120 N.C. at 619, 27 S.E. at 84 (rejecting the defendant's claim that killing chickens out of an "impulse of anger" was legally justified and therefore did not constitute cruelty); State v. Dickens, 215 N.C. 303, 305, 1 S.E.2d 837, 839 (1939) ("The right to slay him cannot be justified by [the dog's] previous act of bursting in through a door, or by the fact that his body emitted an odor peculiar to dogs, but is founded only on the right to protect person or property.").

15. Until 1998 the statute did not specifically state that the act must be without justification. Instead, it provided that the act needed to have been done willfully. N.C. Code § 2482 (1883). Earlier, courts interpreted the term *willfully* to mean more than just intentionally; they required a showing that the act was done "'without just cause, excuse,

vide useful guidance in determining what constitutes a legally defensible justification.

The only other term specifically defined in the misdemeanor cruelty law is *torment,* which refers to "any act, omission or neglect causing or permitting unjustifiable pain, suffering, or death."[16]

Felony Cruelty

The felony animal cruelty law is distinguishable from the misdemeanor in two significant ways. First, it identifies several specific acts of cruelty that are arguably more brutal, such as mutilation and poisoning. Second, it requires a different level of intent: a felony prosecution must show that the person acted maliciously rather than merely intentionally.[17]

Under the statute, it is a Class I felony for a person to maliciously

- torture,
- mutilate,
- maim,
- cruelly beat,
- disfigure,
- poison, or
- kill an animal.[18]

A person may also be charged with a felony if he or she causes or procures an act resulting in one of these seven types of cruelty.[19] The terms *torture* and *cruelly* are synonymous with *torment* as used in the misdemeanor law in that they refer to "any act, omission or neglect causing or permitting unjustifiable pain, suffering, or death."[20]

The statute defines the term *maliciously* to mean that the act is committed not only intentionally but also with malice or bad motive.[21] Given that *malicious* incorporates *intentionally,* the statutory definitions of both terms are relevant when considering the full meaning of *maliciously.* Because *intentionally* is defined as "knowingly and without justifiable excuse," an act of cruelty is

---

or justification.'" *Dickens,* 215 N.C. at 305, 1 S.E.2d at 837 (quoting State v. Yelverton, 196 N.C. 64, 66, 144 S.E. 534, 535 (1928).

16. G.S. 14-360(c). Interestingly, the law applies the same definition to two other terms that are used in the context of felony cruelty: *torture* and *cruelly.*

17. See n. 9 above for the sole use of the term *maliciously* in the misdemeanor statute.

18. Note that a separate law applies if a person injures or kills an animal used for law enforcement purposes. G.S. 14-163.1 (see discussion below).

19. G.S. 360(b)

20. G.S. 14-360(c).

21. *Id.*

malicious if it is committed knowingly (with awareness of or consciousness of what one is doing), without a justifiable excuse, and with malice or bad motive.[22]

It is not entirely clear, however, how the term *malice*, which was added to the statute in 1998, would be interpreted and applied in the context of an animal cruelty case.[23] In homicide cases, North Carolina courts have recognized three meanings for the term:[24]

- the act is done with ill will, hatred, or spite;[25]
- the act that causes death is inherently dangerous to human life and is done so recklessly or wantonly that it reflects disregard of life and social duty; or
- the act is done intentionally and without just cause, excuse, or justification.

These meanings may or may not be appropriate to apply to a felony cruelty case.[26] The third meaning is probably not the most reasonable choice because it overlaps in large part with the statute's definition of *intentionally*—the state of mind required for a charge of misdemeanor cruelty. As discussed above, the definition of *maliciously* in the felony statute incorporates intention and goes further by adding "with malice or bad motive." By including this additional language, the General Assembly probably intended to require different states of mind for the Class 1 misdemeanor versus the felony cruelty.[27]

---

22. *Id.*

23. S.L. 1998-212, sec. 17.16(c).

24. *See* Smith, *North Carolina Crimes*, 4; *see also* State v. Reynolds, 307 N.C. 184, 297 S.E.2d 532 (1982) (discussing the three types of malice recognized in this state).

25. *See* State v. Conrad, 275 N.C. 342, 352, 168 S.E.2d 39, 46 (1969) (explaining that the term *malicious* in the context of a statute criminalizing property damage "connotes a feeling of animosity, hatred or ill will toward the owner, the possessor, or the occupant).

26. In an unpublished animal cruelty opinion, the Court of Appeals appeared to rely primarily on the third meaning when it explained that "malice can be 'the condition of the mind which prompts a person to intentionally inflict serious bodily harm which proximately results in injury without just cause, excuse or justification.'" State v. Dockery, 634 S.E.2d 641, 2006 WL 2671342 (N.C. App. Sept. 19, 2006) (citing State v. Sexton, 357 N.C. 235, 237–38, 581 S.E.2d 57, 58–59 (2003)). While it is important to be aware of this decision, it should not be cited as precedent. According to the North Carolina Rules of Appellate Procedure, "an unpublished decision of the North Carolina Court of Appeals does not constitute controlling legal authority." North Carolina Rules of Appellate Procedure, Rule 30(e), www.aoc.state.nc.us/www/public/html/pdf/ therules.pdf (last visited May 20, 2008).

27. 2A Sutherland Statutory Construction § 46:06 at 181 (6th ed. 2000) ("It is an elementary rule of statutory construction that effect must be given, if possible, to every word, clause and sentence of a statute.").

Exceptions

There are several important exceptions to the misdemeanor and felony cruelty laws. These laws do not apply to

- the taking of animals under the jurisdiction of the Wildlife Resources Commission (WRC), except for those wild birds exempted from the WRC's regulatory definition of "wild birds" (see discussion below),
- activities conducted for the purpose of biomedical research or training,
- activities conducted for the purpose of producing livestock, poultry, or aquatic species,
- activities conducted for the primary purpose of providing food for human or animal consumption,
- activities conducted for veterinary purposes,
- the destruction of any animal for the purposes of protecting the public, other animals, or the public health, and
- the physical alteration of livestock or poultry for the purpose of conforming with breed or show standards.

To be excepted from the criminal law, these six activities must be carried out lawfully. For example, a person who uses an animal for biomedical research in a way not authorized by law may be charged with cruelty.

The language referring to "wild birds" in the wildlife exception has given rise to some confusion in recent years.[28] Before addressing that confusion, it is important to have a clear understanding of the statutory language. Under current wildlife laws, wild (undomesticated) birds that are native to North Carolina are under the WRC's jurisdiction. A WRC regulation exempts four species of nonnative wild birds from its jurisdiction.[29] Because the cruelty statute explicitly provides that any wild bird not included in the WRC's definition

---

28. The cruelty law excepts the "lawful taking of animals under the jurisdiction and regulation of the Wildlife Resources Commission, except that this section shall apply to those birds exempted by the Wildlife Resources Commission from its definition of 'wild birds' pursuant to G.S. 113-129(15a)." G.S. 113-129(15a) defines the term *wild birds* as follows:

> Migratory game birds; upland game birds; and all undomesticated feathered vertebrates. The Wildlife Resources Commission may by regulation list specific birds or classes of birds excluded from the definition of wild birds based upon the need for protection or regulation in the interests of conservation of wildlife resources.

29. N.C. Admin. Code tit 15A, ch. 10B, § .0121 (hereinafter N.C.A.C.). The four exempt species are English sparrow (Passer domesticus), pigeon (Columba livia), mute swan (Cygnus olor), and starling (Sturnus vulgaris). 15A N.C.A.C. 10B .0121.

is protected under the cruelty statute, a person could be charged with criminal cruelty for actions related to birds of the four exempt species. The wild bird exception was litigated for several years in *Malloy v. Cooper,* which involved a biannual pigeon shoot.[30] John Malloy, the plaintiff, sponsored the sporting activity on his property and was concerned that he would be charged with criminal cruelty in connection with the shoot. At the time of the litigation, the WRC exempted only "domestic pigeons," rather than "pigeons," from its jurisdiction.

Malloy asked the court to interpret the law prior to his scheduled pigeon shoot. The Court of Appeals concluded that because domestic and feral pigeons are genetically identical, the cruelty statute was "unconstitutionally vague": people would not know whether they were shooting a domestic pigeon (protected by the cruelty statute) or a feral pigeon (arguably not protected by the cruelty statute). The court explained that the law failed "to give a person a reasonable opportunity to know whether shooting particular pigeons is prohibited, and fail[ed] to provide standards for those applying the law."[31] Because the law was found to be unconstitutional as applied to Malloy's situation, the court stated that it was unenforceable against him.

Shortly after the court issued its decision, the WRC amended its regulation to exempt all pigeons from the definition of "wild bird."[32] As a result, it is clear now that all pigeons, domestic and feral, are protected by the cruelty statute.

Instigating and Promoting Cruelty

Even a person who does not directly hurt an animal may be found criminally responsible for instigating or promoting cruelty. A separate statute makes it a Class 1 misdemeanor to "willfully set on foot, or instigate, or move to, carry on, or promote, or engage in, or do any act towards the furtherance of any act of cruelty to any animal."[33]

While there are no reported North Carolina decisions interpreting this law, courts in several other states have interpreted similar laws. In Arkansas, for example, the court of appeals upheld a woman's conviction for "promoting" dog fighting based on evidence indicating that she was "aware that on property owned by her and her husband an arena had been built for the specific purpose of clandestine dog fighting and that she was aware that it was being so used."[34]

---

30. *Malloy,* 162 N.C.App. 504, 592 S.E.2d 17 (2004).
31. *Id.* at 510, 592 S.E.2d at 22.
32. 18 N.C. Reg. 1598, 1599 (March 15, 2004).
33. G.S. 14-361.
34. Ash v. State, 718 S.W.2d 930, 933 (Ark. 1986).

Reporting Animal Cruelty

North Carolina law does not require any person to report suspected animal cruelty. However, a veterinarian who has reasonable cause to believe that an animal has been subject to cruelty will be protected from civil and criminal liability—as well as any professional disciplinary action—for reporting cruelty, participating in a cruelty investigation, or testifying in cruelty-related judicial proceedings.[35] The state law also protects veterinarians from disciplinary actions by the North Carolina Veterinary Medical Board for failing to report suspected cruelty.

Animal Fighting Exhibitions

In addition to the general cruelty law discussed above, several statutes address specific types of cruelty. Of these statutes, the animal fighting laws are probably used most frequently by local governments. Three separate animal fighting statutes govern cockfighting, dogfighting and baiting, and exhibitions featuring fights between or among all other animals.

North Carolina law does not define the terms *fighting* and *baiting*, but some other jurisdictions do. The District of Columbia, for example, defines *fighting* as "an organized event wherein there is a display of combat between [two] or more animals in which the fighting, killing, maiming, or injuring of an animal is a significant feature, or main purpose, of the event." The term *baiting* means "to attack with violence, to provoke, or to harass an animal with one or more animals for the purpose of training an animal for, or to cause an animal to engage in, fights with or among other animals."[36]

Cockfighting

Under North Carolina the law, it is a Class I felony to be involved in the sport of cockfighting.[37] Specifically, it is illegal for a person to

- instigate, promote, or conduct a cockfight,
- be employed at a cockfight,
- allow property under his or her ownership or control to be used for a cockfight,
- participate as a spectator at a cockfight, or
- profit from a cockfight.

---

35. G.S. 14-360.1. This provision was added to the statute in 2007. S.L. 2007-232.
36. D.C. Code § 22-1015(c). (2001).
37. G.S. 14-362. This law was amended in 2005 to increase the penalty from a misdemeanor to a felony. S.L. 2005-437.

The law further states that a lease of property that is either used for or intended to be used for a cockfighting exhibition is void and that a landlord who learns that the property is being used or will be used for cockfighting must evict the tenant immediately. Some states have also elected to criminalize the ownership of fighting cocks and fighting implements, but North Carolina has not done so.[38]

Dogfighting and Baiting

The dogfighting and dog baiting law is similar to the cockfighting law but goes a bit further. It begins with the same general provisions, making it illegal for a person to

- instigate, promote, or conduct a dogfight,
- be employed at a dogfight,
- allow property under his or her ownership or control to be used for a dogfight,
- participate as a spectator at a dogfight, or
- profit from a dogfight.

The dogfighting and baiting law also makes it illegal for a person to

- provide a dog for a dogfight
- gamble on a dogfight, or
- own, possess, or train a dog with the intent that the dog be used in an exhibition featuring the fighting or baiting of that dog.

The law includes the same language as the cockfighting law in regard to leases of property used for fighting and the duty of a landlord to evict tenants immediately. A violation of the dogfighting and baiting law is a Class H felony, which is one classification higher than cockfighting.

The law was recently amended to address some confusion regarding the scope of the dogfighting law. Language was added to clarify that the law applies to fights between two dogs or between a dog and any other animal.[39] In addition, provisions have been added stating that the dogfighting and baiting law does not prohibit the use of dogs

- for lawful hunting activities governed by the Wildlife Resources Commission,[40]

---

38. *See, e.g.*, Colo. Rev. Stat. Ann. § 18-9-204 (2004); Md. Code Ann., Crim. Law § 10-608 (2004) (criminalizing both the ownership of cocks and implements in Colorado and Maryland). *See also* Humane Society of the United States (HSUS), Cockfighting: State Laws (April 2004) (providing a survey of cockfighting laws in all fifty states), http://files.hsus.org/web-files/PDF/cockfighting_statelaws.pdf (last visited Feb. 20, 2008).

39. S.L. 2006-113, sec. 3.1

40. *Id.*

- as herding dogs engaged in the working of domesticated livestock for agricultural, entertainment, or sporting purposes,[41] or
- in certain earthdog trials.[42]

An earthdog trial is a sporting event in which certain breeds of dogs, specifically terriers and dachshunds, attempt to locate a "quarry" (such as a caged rat) that is in an underground den. According to the American Kennel Club, the trials are designed to test the dog's "natural aptitude and trained hunting and working behaviors when exposed to an underground hunting situation."[43] To be considered exempt from the fighting and baiting law, earthdog trials must be sanctioned or sponsored by an entity approved by the commissioner of agriculture, and the quarry must be kept separate from the dogs by a sturdy barrier and have access to food and water.

Fighting of Other Animals

The third and final criminal fighting statute applies to all animals other than cocks and dogs.[44] This law is virtually identical to the dogfighting and baiting law, with two exceptions. First, it does not specifically prohibit gambling on these animal fighting exhibitions, even though gambling on such exhibitions is illegal under a different criminal statute.[45] Second, the criminal penalties are different. A person who violates this law is guilty of a Class 2 misdemeanor. A subsequent violation for specified acts within three years is a Class I felony.[46]

---

41. S.L. 2007-181.

42. S.L. 2007-180.

43. *See* American Kennel Club, Getting Started in Earthdog Tests, www.akc.org/events/earthdog/getting_started.cfm (last visited Feb. 20, 2008).

44. G.S. 14-362.1.

45. G.S. 14-292 ("[A]ny person or organization that operates any game of chance or any person who plays at or bets on any game of chance at which any money, property or other thing of value is bet, whether the same be in stake or not, shall be guilty of a Class 2 misdemeanor."). Under the law, animal fighting exhibitions are considered "games of chance" rather than games of skill, even though there may be some skill involved on the part of the animals. *See, e.g.*, State v. Brown, 221 N.C. 301, 307, 20 S.E.2d 286, 291 (1942) (concluding that horse racing is a game of chance, regardless of the fact that racing involves skill on the part of the jockey and the horse).

46. It is a felony only if the second offense is for one of the following: instigating, promoting, conducting, being employed at, providing an animal for, or profiting from an animal fighting exhibition. The felony penalty does not apply to owning or possessing an animal, training an animal to fight, or participating as a spectator at an exhibition. G.S. 14-362.1(d).

Spectators at Fighting Exhibitions

All three of the state's animal fighting laws make it a crime to be a spectator at a fighting exhibition. According to the Humane Society of the United States (HSUS), watching a dogfight is a crime in all but four states (Georgia, Idaho, Hawaii, and Montana).[47] HSUS asserts that spectators should be held criminally responsible because they provide the funding, through admission fees and gambling, for the exhibitions.[48]

Recently in North Carolina, a person convicted of being a spectator at a dogfight challenged the constitutionality of this provision, arguing that the state had exceeded the scope of its authority. The Court of Appeals upheld the law, explaining that it is a valid exercise of the state's police power because it is "substantially related" to the object of discouraging dogfighting exhibitions: "If no one attended the dogfights, either for amusement or profit, dogfighting as a group activity would be in jeopardy."[49] One of the three judges dissented from the decision. He agreed with the majority's conclusion that the dogfighting law is constitutional but believed that the state should have offered more evidence establishing that the defendant in this case actually "participated" as a spectator. Specifically, the judge seemed troubled by the fact that the law enforcement official who arrested the spectators testified that "he did not observe whether defendant was actually watching the dogfight."[50]

The chart on p. 13 shows the actions prohibited in the three state laws governing animal fighting exhibitions. The dogfighting law is the most comprehensive, while the cockfighting law is the least comprehensive.

---

47. HSUS, Fact Sheet: Dogfighting: State Laws (March 2007), www.hsus.org/web-files/PDF/dogfighting_ statelaws.pdf (last visited Feb. 20, 2008).

48. HSUS, Dogfighting Fact Sheet, www.hsus.org/acf/fighting/dogfight/dogfighting_fact_sheet.html (last visited March 11, 2008).

49. State v. Arnold, 147 N.C. App. 670, 674, 557 S.E.2d 119, 122 (2001), aff'd 356 N.C. 291, 569 S.E.2d 648 (2002).

50. *Id.* at 676–77, 557 S.E.2d at 123–24 (Wynn, J., dissenting). A similar animal fighting spectator law was challenged in Colorado when a journalist videotaping and reporting on a dogfight was convicted. People v. Bergen, 883 P.2d 532 (Colo. App. 1994). The court rejected the reporter's argument that he should not have been arrested because his journalistic activities were protected by the First Amendment. The court explained that the law did not prevent the reporter from gathering information about dogfighting but did prohibit anyone from attending a dogfight.

A Comparison of North Carolina's Criminal Laws Governing Animal Fighting Exhibitions

	Birds	Dogs	Other
Instigating a fight	☒	☒	☒
Promoting a fight	☒	☒	☒
Conducting a fight	☒	☒	☒
Being employed at a fight	☒	☒	☒
Providing an animal for a fight		☒	☒
Allowing property owned or controlled to be used for a fight	☒	☒	☒
Participating as a spectator at a fight	☒	☒	☒
Gambling on a fight		☒	
Profiting from a fight	☒	☒	☒
Owning, possessing, or training an animal for use in fighting exhibitions		☒	☒

Animal Fighting under Federal Law

At the federal level, the Animal Welfare Act criminalizes various activities related to "animal fighting ventures"—defined as an "event which involves a fight between at least two animals and is conducted for purposes of sport, wagering, or entertainment."[51] The federal law generally supplements state laws governing fighting and only overrides a state or local animal fighting law if it is in "direct and irreconcilable conflict" with the federal statute.[52]

Four general categories of activities are prohibited under the federal law:

- Sponsoring or exhibiting an animal in an animal fighting venture when any animal in the venture has been moved via interstate or foreign commerce.[53]
- Buying, selling, delivering, or transporting an animal via interstate or foreign commerce for the purpose of having it participate in an animal fighting venture.[54]

---

51. 7 U.S.C. § 2156(g)(1). Hunting-related activities are excluded from the definition.

52. 7 U.S.C. § 2156(h) ("The provisions of this chapter shall not supersede or otherwise invalidate any such State, local, or municipal legislation or ordinance relating to animal fighting ventures except in case of a direct and irreconcilable conflict between any requirements thereunder and this chapter or any rule, regulation, or standard hereunder.").

53. 7 U.S.C. § 2156(a)(1). There is a narrow exception to the law that applies to fighting ventures involving live birds (i.e., cockfights) in states where the venture is legal.

54. *Id.* at § 2156(b). The law also applies if a person "receives" an animal for the purpose of transporting it to another another state or country where it will be used in an animal fighting venture.

- Using the mail service or any instrumentality of interstate commerce to promote or in any other manner further an animal fighting venture.[55]
- Buying, selling, delivering, or transporting in interstate or foreign commerce a knife, gaff, or any other sharp instrument attached, or designed to be attached, to the leg of a bird for use in an animal fighting venture.[56]

Because of constitutional limitations on federal authority, these provisions all relate to the transport of the animals, equipment, or information through interstate or foreign commerce.[57] Thus, a fighting venture or sale of cockfighting implements that is wholly intrastate would not be subject to federal law; that is, if the animal fight took place in a single state, the animals were not transported across state lines, and no communication about the fight was sent across state lines through the mail or other means.

This federal law is enforced by the Animal and Plant Health Inspection Service and the Office of the Inspector General of the U.S. Department of Agriculture.[58] For each violation, a person found guilty may be imprisoned for up to three years, fined, or both fined and imprisoned.[59]

---

55. *Id.* at § 2156(c). This portion of the law does not apply when the conduct is performed "outside the limits of the States of the United States."

56. *Id.* at § 2156(e). This language was added by legislation that passed Congress in April 2007. Pub. L. No. 110-22 (to be codified at 18 U.S.C. § 49 and 7 U.S.C. § 2156).

57. *See e.g.,* Slavin v. U.S., 403 F.3d 522 (8th Cir. 2005) (upholding the statute as a constitutional exercise of federal authority to regulate interstate commerce).

58. Press Release, U.S. Department of Agriculture, The Animal Welfare Act Provisions on Animal Fighting (Aug. 2003), www.aphis.usda.gov/lpa/pubs/fsheet_faq_notice/fs_awafighting.html (last visited Feb. 22, 2008). The Department of Agriculture has been criticized in the past for failing to adequately enforce the Animal Welfare Act. *See, e.g.,* Shigehiko Ito, "Beyond Standing: A Search for a New Solution in Animal Welfare," *Santa Clara Law Review* 46, no. 2 (2006): 377, 378 .

59. Pub. L. No. 110-22 (to be codified at 18 U.S.C. § 49. The penalty was increased from a misdemeanor to a felony in April 2007. Animal Fighting Prohibition Enforcement Act of 2007, H.R. 137, 110th Cong. (2007). The House committee report endorsing the legislation explained that increasing the penalty to a felony would lead to more prosecutions. H.R. Rep. No 110-27, pt. 1, at 2 (2007) ("Prohibitions against knowingly selling, buying, transporting, delivering, or receiving an animal in interstate or foreign commerce for the purposes of participation in an animal fighting venture were added to the Animal Welfare Act in 1976, with misdemeanor penalties of up to $5,000 in fines and up to 1 year in prison. Since then, Federal authorities have pursued fewer than a half dozen animal fighting cases, despite receiving numerous tips from informants and requests to assist with state and local prosecutions. The animal fighting industry continues to thrive within the United States, despite 50 State laws that ban dogfighting and 48 State laws that ban cockfighting. . . . By increasing penalties to the felony level, H.R. 137

Other Criminal Laws

State and federal law also criminalize other specific activities that involve cruelty or mistreatment of animals. Under federal law, for example, it is a crime to create, sell, or possess a depiction of animal cruelty with the intention of placing the depiction in interstate or foreign commerce for commercial gain.[60]

Poison Control

In North Carolina three separate statutes address the poisoning of animals. One makes it unlawful to throw or leave a poisonous shrub, plant, tree, or vegetable exposed on any city street, alley, or open lot or on a public road anywhere in the state.[61] Another prohibits placing strychnine, other poisonous compounds, or ground glass on any food left in several specific open areas where animals might roam. The same law also prohibits leaving open containers of antifreeze in those same open areas.[62] A violation of either of these two laws is a misdemeanor. The third poisoning statute makes it a felony to poison livestock.[63]

Law Enforcement and Assistance Animals

Law enforcement and assistance animals have special protections under both state and federal law.[64] Under state law, a person who knows or has reason to

---

will give prosecutors greater incentive to pursue cases against unlawful animal fighting ventures, and strengthen deterrence against them.").

60. 18 U.S.C. § 48(a). The law excepts depictions that have "serious religious, political, scientific, educational, journalistic, historical, or artistic value." *Id.* at § 48(b). A *depiction* is defined as "any visual or auditory depiction, including any photograph, motion-picture film, video recording, electronic image, or sound recording of conduct in which a living animal is intentionally maimed, mutilated, tortured, wounded, or killed, if such conduct is illegal under Federal law or the law of the state." *Id.* at § 48(c)(1).

61. G.S. 14-368. Any person committing this offense is liable for civil damages and is also guilty of a Class 2 misdemeanor.

62. G.S. 14-401. These substances must not be placed in "any public square, street, lane, alley or on any lot in any village, town or city or on any public road, open field, woods, or yard in the country." Violators may be liable for civil damages and found guilty of a Class 1 misdemeanor.

63. G.S. 14-163.

64. G.S. 14-163.1. The term *law enforcement agency animal* is defined as "an animal that is trained and may be used to assist a law enforcement officer in the performance of the officer's official duties." G.S. 14-163.1(a)(2). An *assistance animal* is "an animal that is trained and may be used to assist a 'person with a disability' as defined in G.S. 168A-3." G.S. 14-163.1(a)(1). A "person with a disability" is "any person who (i) has a physical or mental impairment which substantially limits one or more major life activities; (ii) has a record of such an impairment; or (iii) is regarded as having such an impairment." G.S. 168A-3(7a).

know that an animal is a law enforcement agency animal or an assistance animal and

- willfully causes or attempts to cause serious harm to the animal is guilty of a Class I felony.[65]
- willfully causes or attempts to cause harm to the animal is guilty of a Class 1 misdemeanor.[66]
- willfully taunts, teases, harasses, delays, obstructs or attempts to delay or obstruct the animal in the performance of its duty is guilty of a Class 2 misdemeanor.[67]

Under this law, the term *harm* means "any injury, illness, or other physiological impairment; or any behavioral impairment that impedes or interferes with duties performed by" the animal.[68] The term *serious harm*, which is used in the context of the felony, is defined to include any harm that

- creates a substantial risk of death,
- causes maiming or substantial loss or impairment of bodily function,
- causes acute pain of a duration that results in substantial suffering,
- requires retraining of the animal, or
- requires retirement of the animal.[69]

In June 2007 the General Assembly enacted legislation to amend this law to also make it a Class H felony to willfully kill or attempt to kill a law enforcement or assistance animal.[70] Under federal law, it is a crime to willfully and maliciously harm a dog or horse used by a federal agency to enforce the law, detect criminal activity, or apprehend criminals.[71]

Other Cruelty-Related Misdemeanors

Abandonment. It is a Class 2 misdemeanor In North Carolina for a person who owns, possesses, or has charge or custody of an animal to willfully abandon it without a justifiable excuse.[72]

---

65. G.S. 14-163.1(b).
66. G.S. 14-163.1(c).
67. G.S. 14-163.1(d).
68. G.S. 163.1(a)(3).
69. G.S. 14-163.1(a)(4).
70. S.L. 2007–80. The legislation also allows a court to consider as an aggravating factor for sentencing purposes evidence indicating that the animal seriously harmed or killed was engaged in the performance of its official duties. *Id.* at sec. 2 (amending G.S. 15A-1340.16).
71. 18 U.S.C. § 1368.
72. G.S. 14-361.1.

Unlawful restraint. A person will be guilty of a Class 1 misdemeanor if he or she maliciously restrains a dog using a chain or wire that is much larger or heavier than is needed to restrain the dog safely.[73] In the context of this law, the term *maliciously* means that the person used the restraint (1) intentionally and (2) with malice or bad motive.

Conveying animals. It is a Class 1 misdemeanor to convey an animal in or upon a vehicle or other conveyance in a cruel or inhuman manner.[74] The law provides that when someone is taken into custody for a violation of this law, the officer has the authority to take charge of the conveyance and take steps to recover the costs of maintaining it while the person is in custody.[75]

Disposition of certain young animals. It is a Class 3 misdemeanor in North Carolina to sell, offer for sale, barter, or give away as premiums (or prizes) certain young animals as pets or novelties. The law applies to chicks, ducklings, or other fowl, and rabbits under eight weeks of age.[76]

Often when a person is charged with a cruelty-related offense, law enforcement or animal control officials seize the animals as evidence or to protect them. As a result, the local government may incur significant expenses in providing care and shelter for the seized animals. State law allows certain groups—including animal shelters operated by or under contract with a local government—to recover some of these costs.[77] For a full discussion of this cost-recovery mechanism, see the discussion in Chapter 2 beginning on page 36.

Local Laws

Local governments have long had the authority to regulate the treatment of animals. Cities and counties have specific statutory authority to "define and prohibit the abuse of animals."[78] They can also "define, prohibit, regulate, or abate acts, omissions, or conditions, detrimental to the health, safety, or welfare of its citizens and the peace and dignity of the city, and may define and abate nuisances."[79] The combination of these two statutory grants of authority provides local governments with relatively broad authority in this field.

---

73. G.S. 14-362.3.

74. G.S. 14-363.

75. Specifically, the law allows the officer to incur expenses necessary to keep and sustain the vehicle and to impose a lien on the vehicle that the defendant must pay before reclaiming the vehicle.

76. G.S. 14-363.1.

77. G.S. 19A-70.

78. *See* G.S. 153A-127 (counties); 160A-182 (cities).

79. G.S. 160A-174(a).

This authority is not, however, without limits. Specifically, an ordinance must not

- infringe a liberty guaranteed to the people by the state or federal constitution;
- make unlawful an act, omission, or condition that is expressly made lawful by state or federal law;
- make lawful an act, omission, or condition expressly made unlawful by state or federal law;
- purport to regulate a subject that local governments are expressly forbidden to regulate by state or federal law;
- purport to regulate a field that a state or federal statute clearly reflects a legislative intention to provide a complete and integrated regulatory scheme exclusive of local regulation; or
- define the elements of an offense in a way that is identical to the elements of an offense defined by state or federal law.[80]

In short, a local ordinance may regulate the same conduct as a state or federal law, but it must not duplicate or undermine the other law. Rather, it should impose higher standards or expectations within the jurisdiction. Given the broad scope of state and federal cruelty-related laws, local government officials drafting a local ordinance need to be familiar with these laws to ensure that they do not run afoul of the restrictions described above.[81] An example of a local ordinance that appropriately builds on state law is Asheville's ordinance that prohibits leaving animals in motor vehicles under conditions that would endanger their health or well-being.[82] While leaving an animal confined in a hot car could be considered cruelty under the state's general cruelty statute, sin-

80. G.S. 160A-174(b). While these limitations are named only in the law governing municipalities, the courts have consistently applied them to counties as well. *See* State v. Tenore, 280 N.C. 238, 248, 185 S.E.2d 644, 650 (1972).

81. *See* G.S. 153A-121 (general ordinance making power of counties); G.S. 160A-174 (general ordinance making power of cities).

82. Asheville Code of Ordinances, § 2-12(e) ("It shall be unlawful for any person to place or confine an animal or allow an animal to be placed or confined in a motor vehicle under such conditions or for such a period of time as to endanger the health or well-being of such animal due to temperature, lack of food or drink, or such other conditions as may reasonably be expected to cause suffering, disability, or death. After making a reasonable effort to find the driver of a vehicle in which an animal is confined, the animal control officer, in the presence of a law enforcement officer, may use the least intrusive means to enter the vehicle if necessary to remove the animal, where reasonable cause exists to believe the animal may die if not immediately removed.").

gling out this action in a local ordinance is a reasonable means of addressing a specific local concern without risking explicit duplication of state law.

In some jurisdictions, board of health rules govern animal control and may include provisions related to cruelty or abuse. A board of health is most likely to become involved in animal control if the local health department is the agency with administrative responsibility for the county's animal control program. Yet, even though boards of health may play some oversight role of animal control activities by virtue of the health department's administrative role, it is not clear that they have the legal authority to adopt comprehensive animal control rules. Under state law, the rule-making authority of boards of health is limited to rules necessary to "protect and promote the public health." The term *public health* usually refers to issues affecting *human* health.[83] Therefore, while a board of health may appropriately adopt a rule governing rabies, it may not be appropriate for the board to adopt rules on animal issues unrelated to human health, such as cruelty or nuisance animals.

Conclusion

All three levels of government—federal, state, and local—address animal cruelty in different contexts and assign different penalties. The criminal laws discussed above provide several possible avenues for responding to alleged cruelty, while the civil remedies discussed in the next chapter offer individuals and government officials an entirely different remedy—a civil injunction. Both the civil and criminal remedies should be considered when evaluating the appropriate response to an act of animal cruelty.

---

83. G.S. 130A-39(a). *See, e.g.,* Institute of Medicine, National Academy of Sciences, *The Future of Public Health* (Washington D.C.: National Academy Press, 1988) (characterizing public health's mission as "fulfilling society's interest in assuring conditions in which people can be healthy").

Relevant Statutes

Article 23 of Chapter 14
Trespasses to Personal Property.

. . .

§ 14-163. Poisoning livestock.
If any person shall willfully and unlawfully poison any horse, mule, hog, sheep or other livestock, the property of another, such person shall be punished as a Class I felon.

§ 14-163.1. Assaulting a law enforcement agency animal or an assistance animal.

(a) The following definitions apply in this section:

(1) Assistance animal. – An animal that is trained and may be used to assist a "person with a disability" as defined in G.S. 168A-3. The term "assistance animal" is not limited to a dog and includes any animal trained to assist a person with a disability as provided in Article 1 of Chapter 168 of the General Statutes.

(2) Law enforcement agency animal. – An animal that is trained and may be used to assist a law enforcement officer in the performance of the officer's official duties.

(3) Harm. – Any injury, illness, or other physiological impairment; or any behavioral impairment that impedes or interferes with duties performed by a law enforcement agency animal or an assistance animal.

(4) Serious harm. – Harm that does any of the following:

a. Creates a substantial risk of death.

b. Causes maiming or causes substantial loss or impairment of bodily function.

c. Causes acute pain of a duration that results in substantial suffering.

d. Requires retraining of the law enforcement agency animal or assistance animal.

e. Requires retirement of the law enforcement agency animal or assistance animal from performing duties.

(b) Any person who knows or has reason to know that an animal is a law enforcement agency animal or an assistance animal and who willfully causes or attempts to cause serious harm to the animal is guilty of a Class I felony.

(c) Unless the conduct is covered under some other provision of law providing greater punishment, any person who knows or has reason to know that an animal is a law enforcement agency animal or an assistance animal and who willfully causes or attempts to cause harm to the animal is guilty of a Class 1 misdemeanor.

(d) Unless the conduct is covered under some other provision of law providing greater punishment, any person who knows or has reason to know that an animal is a law enforcement agency animal or an assistance animal and who willfully taunts, teases, harasses, delays, obstructs, or attempts to delay or obstruct the animal in the performance of its duty as a law enforcement agency animal or assistance animal is guilty of a Class 2 misdemeanor.

(d1) A defendant convicted of a violation of this section shall be ordered to make restitution to the person with a disability, or to a person, group, or law enforcement agency who owns or is responsible for the care of the law enforcement agency animal for any of the following as appropriate:

(1) Veterinary, medical care, and boarding expenses for the assistance animal or law enforcement animal.
(2) Medical expenses for the person with the disability relating to the harm inflicted upon the assistance animal.
(3) Replacement and training or retraining expenses for the assistance animal or law enforcement animal.
(4) Expenses incurred to provide temporary mobility services to the person with a disability.
(5) Wages or income lost while the person with a disability is with the assistance animal receiving training or retraining.
(6) The salary of the law enforcement agency animal handler as a result of the lost services to the agency during the time the handler is with the law enforcement agency animal receiving training or retraining.
(7) Any other expense reasonably incurred as a result of the offense.

(e) This section shall not apply to a licensed veterinarian whose conduct is in accordance with Article 11 of Chapter 90 of the General Statutes.

(f) Self-defense is an affirmative defense to a violation of this section.

(g) Nothing in this section shall affect any civil remedies available for violation of this section.

Article 47 of Chapter 14
Cruelty to Animals.

§ 14-360. Cruelty to animals; construction of section.

(a) If any person shall intentionally overdrive, overload, wound, injure, torment, kill, or deprive of necessary sustenance, or cause or procure to be overdriven, overloaded, wounded, injured, tormented, killed, or deprived of necessary sustenance, any animal, every such offender shall for every such offense be guilty of a Class 1 misdemeanor.

(a1) If any person shall maliciously kill, or cause or procure to be killed, any animal by intentional deprivation of necessary sustenance, that person shall be guilty of a Class A1 misdemeanor.

(b) If any person shall maliciously torture, mutilate, maim, cruelly beat, disfigure, poison, or kill, or cause or procure to be tortured, mutilated, maimed, cruelly beaten, disfigured, poisoned, or killed, any animal, every such offender shall for every such offense be guilty of a Class I felony. However, nothing in this section shall be construed to increase the penalty for cockfighting provided for in G.S. 14-362.

(c) As used in this section, the words "torture", "torment", and "cruelly" include or refer to any act, omission, or neglect causing or permitting unjustifiable pain, suffering, or death. As used in this section, the word "intentionally" refers to an act committed knowingly and without justifiable excuse, while the word "maliciously" means an act committed intentionally and with malice or bad motive. As used in this section, the term "animal" includes every living vertebrate in the classes Amphibia, Reptilia, Aves, and Mammalia except human beings. However, this section shall not apply to the following activities:

(1) The lawful taking of animals under the jurisdiction and regulation of the Wildlife Resources Commission, except that this section shall apply to those birds exempted by the Wildlife Resources Commission from its definition of "wild birds" pursuant to G.S. 113-129(15a).

(2) Lawful activities conducted for purposes of biomedical research or training or for purposes of production of livestock, poultry, or aquatic species.

(2a) Lawful activities conducted for the primary purpose of providing food for human or animal consumption.

(3) Activities conducted for lawful veterinary purposes.

(4) The lawful destruction of any animal for the purposes of protecting the public, other animals, property, or the public health.

(5) The physical alteration of livestock or poultry for the purpose of conforming with breed or show standards.

§ 14-360.1. Immunity for veterinarian reporting animal cruelty.
Any veterinarian licensed in this State who has reasonable cause to believe that an animal has been the subject of animal cruelty in violation of G.S. 14-360 and who makes a report of animal cruelty, or who participates in any investigation or testifies in any judicial proceeding that arises from a report of animal cruelty, shall be immune from civil liability, criminal liability, and liability from professional disciplinary action and shall not be in breach of any veterinarian-patient confidentiality, unless the veterinarian acted in bad faith or with a malicious purpose. It shall be a rebuttable presumption that the veterinarian acted in good faith. A failure by a veterinarian to make a report of animal cruelty shall not constitute grounds for disciplinary action under G.S. 90-187.8."

§ 14-361. Instigating or promoting cruelty to animals.
If any person shall willfully set on foot, or instigate, or move to, carry on, or promote, or engage in, or do any act towards the furtherance of any act of cruelty to any animal, he shall be guilty of a Class 1 misdemeanor.

§ 14-361.1. Abandonment of animals.
Any person being the owner or possessor, or having charge or custody of an animal, who willfully and without justifiable excuse abandons the animal is guilty of a Class 2 misdemeanor.

§ 14-362. Cockfighting.
A person who instigates, promotes, conducts, is employed at, allows property under his ownership or control to be used for, participates as a spectator at, or profits from an exhibition featuring the fighting of a cock is guilty of a Class I felony. A lease of property that is used or is intended to be used for an exhibition featuring the fighting of a cock is void, and a lessor who knows this use is made or is intended to be made of his property is under a duty to evict the lessee immediately.

§ 14-362.1. Animal fights and baiting, other than cock fights, dog fights and dog baiting.

(a) A person who instigates, promotes, conducts, is employed at, provides an animal for, allows property under his ownership or control to be used for, or profits from an exhibition featuring the fighting or baiting of an animal, other than a cock or a dog, is guilty of a Class 2 misdemeanor. A lease of property that is used or is intended to be used for an exhibition featuring the fighting or baiting of an animal, other than a cock or a dog, is void, and a lessor who knows this use is made or is intended to be made of his property is under a duty to evict the lessee immediately.

(b) A person who owns, possesses, or trains an animal, other than a cock or a dog, with the intent that the animal be used in an exhibition featuring the fighting or baiting of that animal or any other animal is guilty of a Class 2 misdemeanor.

(c) A person who participates as a spectator at an exhibition featuring the fighting or baiting of an animal, other than a cock or a dog, is guilty of a Class 2 misdemeanor.

(d) A person who commits an offense under subsection (a) within three years after being convicted of an offense under this section is guilty of a Class I felony.

(e) This section does not prohibit the lawful taking or training of animals under the jurisdiction and regulation of the Wildlife Resources Commission.

§ 14-362.2. Dog fighting and baiting.

(a) A person who instigates, promotes, conducts, is employed at, provides a dog for, allows property under the person's ownership or control to be used for, gambles on, or profits from an exhibition featuring the baiting of a dog or the fighting of a dog with another dog or with another animal is guilty of a Class H felony. A lease of property that is used or is intended to be used for an exhibition featuring the baiting of a dog or the fighting of a dog with another dog or with another animal is void, and a lessor who knows this use is made or is intended to be made of the lessor's property is under a duty to evict the lessee immediately.

(b) A person who owns, possesses, or trains a dog with the intent that the dog be used in an exhibition featuring the baiting of that dog or the fighting of that dog with another dog or with another animal is guilty of a Class H felony.

(c) A person who participates as a spectator at an exhibition featuring the baiting of a dog or the fighting of a dog with another dog or with another animal is guilty of a Class H felony.

(d) This section does not prohibit the use of dogs in the lawful taking of animals under the jurisdiction and regulation of the Wildlife Resources Commission.

(e) This section does not prohibit the use of dogs in earthdog trials that are sanctioned or sponsored by entities approved by the Commissioner of Agriculture that meet standards that protect the health and safety of the dogs. Quarry at an earthdog trial shall at all times be kept separate from the dogs by a sturdy barrier, such as a cage, and have access to food and water.

(f) This section does not apply to the use of herding dogs engaged in the working of domesticated livestock for agricultural, entertainment, or sporting purposes.

§ 14-362.3. Restraining dogs in a cruel manner.

A person who maliciously restrains a dog using a chain or wire grossly in excess of the size necessary to restrain the dog safely is guilty of a Class 1 misdemeanor. For purposes of this section, "maliciously" means the person imposed the restraint intentionally and with malice or bad motive.

§ 14-363. Conveying animals in a cruel manner.

If any person shall carry or cause to be carried in or upon any vehicle or other conveyance, any animal in a cruel or inhuman manner, he shall be guilty of a Class 1 misdemeanor. Whenever an offender shall be taken into custody therefor by any officer, the officer may take charge of such vehicle or other conveyance and its contents, and deposit the same in some safe place of custody. The necessary expenses which may be incurred for taking charge of and keeping and sustaining the vehicle or other conveyance shall be a lien thereon, to be paid before the same can be lawfully reclaimed; or the said expenses, or any part thereof remaining unpaid, may be recovered by the person incurring the same of the owner of such animal in an action therefor.

§ 14-363.1. Living baby chicks or other fowl, or rabbits under eight weeks of age; disposing of as pets or novelties forbidden.

If any person, firm or corporation shall sell, or offer for sale, barter or give away as premiums living baby chicks, ducklings, or other fowl or rabbits under eight weeks of age as pets or novelties, such person, firm or corporation shall be guilty of a Class 3 misdemeanor. Provided, that nothing contained in this section shall be construed to prohibit the sale of nondomesticated species of chicks, ducklings, or other fowl, or of other fowl from proper brooder facilities by hatcheries or stores engaged in the business of selling them for purposes other than for pets or novelties.

§ 14-363.2. Confiscation of cruelly treated animals.

Conviction of any offense contained in this Article may result in confiscation of cruelly treated animals belonging to the accused and it shall be proper for the court in its discretion to order a final determination of the custody of the confiscated animals.

---

§ 14-368. Placing poisonous shrubs and vegetables in public places.
If any person shall throw into or leave exposed in any public square, street, lane, alley or open lot in any city, town or village, or in any public road, any mock orange or other poisonous shrub, plant, tree or vegetable, he shall be liable in damages to any person injured thereby and shall also be guilty of a Class 2 misdemeanor.

§ 14-401. Putting poisonous foodstuffs, antifreeze, etc., in certain public places, prohibited.
It shall be unlawful for any person, firm or corporation to put or place (i) any strychnine, other poisonous compounds or ground glass on any beef or other foodstuffs of any kind, or (ii) any antifreeze that contains ethylene glycol and is not in a closed container, in any public square, street, lane, alley or on any lot in any village, town or city or on any public road, open field, woods or yard in the country. Any person, firm or corporation who violates the provisions of this section shall be liable in damages to the person injured thereby and also shall be guilty of a Class 1 misdemeanor. This section shall not apply to the poisoning of insects or worms for the purpose of protecting crops or gardens by spraying plants, crops, or trees, to poisons used in rat extermination, or to the accidental release of antifreeze containing ethylene glycol.

§ 153A-127. Abuse of animals.
A county may by ordinance define and prohibit the abuse of animals.

§ 160A-182. Abuse of animals.
A city may by ordinance define and prohibit the abuse of animals.

Chapter 2

Civil Cruelty

Besides the criminal statutes governing animal cruelty and animal fighting exhibitions reviewed in Chapter 1, North Carolina law provides civil remedies for protecting animals from persons who abuse, neglect, or otherwise treat them cruelly. This chapter addresses those civil remedies. It first summarizes the process through which any person, regardless of his or her relationship to an animal, may ask a court to order another person to stop treating an animal cruelly. Next, it addresses the laws governing animal cruelty investigators. Finally, the chapter briefly reviews three mechanisms available for recovering some of the costs plaintiffs might incur for the shelter and care of animals taken from their owners while civil cruelty cases are pending.

Civil Cruelty Actions under State Law

North Carolina state law establishes a civil process that allows a court to impose the restrictions it deems necessary to protect an animal that is being treated cruelly.[1] The statute, found in Chapter 19A, Article 1 of the North Carolina General Statutes, is entitled "Civil Remedy for the Protection of Animals." In general, this civil remedy is designed to stop someone from treating an animal cruelly. It is not designed to compensate owners financially for losses related to an animal's injury or death. For example, if Andy is upset with Bob because Bob injured or killed Andy's prize-winning pet, Andy may want to pursue criminal cruelty charges against Bob to see that he is punished; or perhaps he may want to file a civil tort claim against Bob to recover money damages. The civil remedy discussed in this chapter provides for neither of these actions. Rather, its primary purpose is to provide for *injunctions*—court orders "prohibiting someone from doing some specified act or commanding someone to undo some wrong or injury."[2]

---

1. Article 1 of N.C. GEN. STAT. Chapter 19A (hereinafter G.S.).
2. *Black's Law Dictionary,* 8th ed. (Minneapolis: West, 2004), 800.

The civil remedy is available to protect any animal. The term *animal* as defined in the statute includes "every living vertebrate in the classes Amphibia, Reptilia, Aves, and Mammalia except human beings."[3] The term *cruelty* includes "every act, omission, or neglect whereby unjustifiable physical pain, suffering, or death is caused or permitted."[4] These two definitions mirror those used in the criminal cruelty laws.[5]

Standing

In general, only certain classes of individuals are allowed to bring a civil lawsuit. These classes are said to have standing to bring a legal action. According to some animal rights advocates, the issue of standing is often argued as the first round in any cruelty litigation.[6] In North Carolina, though, standing is not an issue in civil animal cruelty cases. The state's law allows *any* person to bring a civil action for animal cruelty, "even though the person does not have a possessory or ownership right in an animal."[7] The term *person* includes political and corporate bodies as well as individuals.[8] Thus an animal protection society, local government, neighbor, or perfect stranger can bring a private lawsuit alleging animal cruelty.[9]

---

3. G.S. 19A(1). One scholar recommends expanding the definition to include the class *Pisces* (fish). *See* William A. Reppy Jr., "Citizen Standing to Enforce Anti-Cruelty Laws by Obtaining Injunctions: The North Carolina Experience," *Animal Law* 11 (2005): 45.

4. G.S. 19A(2).

5. G.S. 14-360(c) (definitions of *cruelly* and *animal*); *see also* Wall, *Animal Cruelty, Part 1,* 2, n. 5 (evolution of the term *animal*). [Or cross-reference to chap. 1 of present work instead?]

6. *See* Delcianna J. Winders, "Confronting Barriers to the Courtroom for Animal Advocates," *Animal Law* 13, no.1 (2006–2007): 6 (quoting Joyce Tischler, co-founder of the Animal Legal Defense Fund: 'We didn't set out to make standing law. We didn't want to become standing experts. Dealing with the issue of standing . . . has become a practical necessity, because we are challenged in every case we file."). For decisions addressing the issue of standing in cruelty cases, *see* American Soc'y for Prevention of Cruelty to Animals v. Ringling Bros. & Barnum & Bailey, 317 F.3d 334, 338 (D.C. Cir. 2003) (holding that a circus animal handler had standing); Animal Legal Def. Fund v. Glickman, 154 F.3d 426, 445 (D.C. Cir. 1998) (recognizing that a visitor to a zoo had standing).

7. G.S. 19A-2.

8. G.S. 19A-1(3). The statute cites the definition of *person* in G.S. 12-3, which refers to "bodies politic and corporate, as well as to individuals, unless the context clearly shows to the contrary." *See* Reppy, "Citizen Standing," 41–44 (discussing the broad standing provisions and explaining that the statute was amended in 2003 to clarify that local governments have standing).

9. Before the civil cruelty law was adopted, the court was unwilling to issue an injunction to prevent actions that may have constituted animal cruelty under the existing criminal law. In 1962, for example, the North Carolina Supreme Court rejected a

Process

The law provides two basic tools for addressing cruelty—a preliminary injunction and a permanent injunction. Plaintiffs typically request a preliminary injunction and then return to court later to request a permanent injunction. The two primary players in these actions are the plaintiff and the defendant. The plaintiff is the person who files the action alleging that another person is treating an animal cruelly. The defendant in a civil cruelty case may be any person who either owns or has possession of the animal.[10]

Preliminary Injunction

To obtain a preliminary injunction, the plaintiff must first file a verified complaint in the district court in the county where the alleged cruelty occurred. The law requires that the complaint be verified, meaning that the person filing the complaint must also file an affidavit. An affidavit is a sworn statement—a written document clearly explaining the facts supporting the request for an injunction that is signed by the person making the statement (the *affiant)* and notarized.[11]

Under the state's rules of civil procedure, the defendant must be served with notice that the complaint has been filed.[12] A court may decide to issue a preliminary injunction if, based only upon the complaint, it appears that the plaintiff is entitled to relief.[13]

Plaintiffs usually have to post a bond sufficient to cover any costs a defendant might incur before a court will issue a preliminary injunction. The bond will be used to reimburse the defendant if the court later determines that the

---

plaintiff's request to enjoin a rabbit hunt allegedly conducted in a cruel manner. Yandell v. Am. Legion Post No. 113, 256 N.C. 691, 693, 124 S.E.2d 885, 886–87 (1962). The court explained that "ordinarily the violation of a criminal statute is not sufficient to invoke equitable jurisdiction of the court." *Id.*

10. G.S. 19A-2. The term *person* is defined broadly to include both individuals and political and corporate bodies. G.S. 12-3. A defendant could be, for example, a private individual, an animal shelter, a circus company, or a pet store.

11. G.S. 1A-1, Rule 11(b)

12. G.S. 1A-1, Rule 65(a). At one time, the cruelty law allowed for a temporary restraining order, which in some cases authorized the court to issue a temporary order before the defendant received notice of the suit. That language was removed in 1979. *See* Reppy, "Citizen Standing," 51–52 (arguing that the language authorizing a temporary restraining order in cruelty cases should be restored).

13. G.S. 1-485. The statute cited also outlines two other circumstances in which a preliminary injunction may be granted, but only the granting of relief that "consists in restraining the commission or continuance of some act" appears relevant to cruelty cases.

injunction was improper.[14] The court sets the amount of the bond or, in some situations, may conclude that no bond is required.[15] If the case is initiated by an officially appointed animal cruelty investigator[16] or by the state or a local government, no bond will be required. The court could, however, later assess damages against a government plaintiff if, for example, it decides not to issue a permanent injunction.[17]

The plaintiff may request permission to take custody of the animal and provide for its care. The court has the discretion to issue such an order if it concludes, based on the plaintiff's complaint, that the "condition giving rise to the cruel treatment . . . requires the animal to be removed" from the defendant's custody.[18] If temporary custody is awarded to the plaintiff, the plaintiff is allowed to place the animal with a foster care provider.[19]

While the animal is in the plaintiff's temporary custody, the plaintiff may decide that the animal needs veterinary care. The law provides the plaintiff with clear authority to obtain any such care (except euthanasia).[20] Before seeking veterinary care, however, the plaintiff is required to consult with or attempt to consult with the defendant about the care. Note that the law does not require the plaintiff to obtain the defendant's permission to seek veterinary care; it only requires that the plaintiff consult, or attempt to consult, with the defendant. Even if the defendant disagrees with the plaintiff's decision, the plaintiff may provide care for the animal.

If the plaintiff concludes that the animal should be euthanized, he or she must obtain either the written consent of the defendant or a court order. The court may issue such an order if it finds that the animal is suffering as a result of an incurable illness or irreparable injury.

---

14. G.S. 1A-1, Rule 65(c) ("No restraining order or preliminary injunction shall issue except upon the giving of security by the applicant, in such sum as the judge deems proper, for the payment of such costs and damages as may be incurred or suffered by any party who is found to have been wrongfully enjoined or restrained.").

15. *See* Keith v. Day, 60 N.C. App. 559, 561–62, 299 S.E.2d 296, 297–98 (1983).

16. G.S. 19A-45(c). For a discussion of the role of animal cruelty investigators, see pp. 33–35 below.

17. G.S. 1A-1, Rule 65(c) ("No such security shall be required of the State of North Carolina or of any county or municipality thereof, or any officer or agency thereof acting in an official capacity, but damages may be awarded against such a party in accord with this rule.").

18. G.S. 19A-3(a).

19. G.S. 19A-3(c).

20. G.S. 19A-3(b).

Permanent Injunction

After a preliminary injunction is issued, most plaintiffs seek a permanent injunction.[21] A district court judge will provide an opportunity for both the plaintiff and the defendant to offer evidence and may then issue a permanent injunction.[22]

If the judge decides not to issue a permanent injunction, the plaintiff's action will be dismissed and the preliminary injunction will be dissolved. If the animal is in the care of a custodian, the judge will probably direct the custodian to return it to the defendant.[23] The judge has the option, however, of extending the alternative custody and care arrangements until the time for appeal expires or until all appeals have been exhausted.[24]

If the judge does issue the permanent injunction but decides that the animal can be safely returned to the defendant, the order will outline the restrictions placed on the defendant.[25] For example, if a dog was kept outside without shelter from the elements, the judge could order the defendant to provide it with appropriate shelter.

The judge may also conclude that there would be a "substantial risk that the animal would be subjected to further cruelty if returned to the possession of the defendant."[26] If so, the judge may terminate the defendant's ownership and right of possession, which means that the defendant would no longer have a right to own or keep the animal. The judge could then transfer the ownership and right of possession to another person or entity, such as the plaintiff or a foster care provider. The judge might also impose restrictions on the defendant's ability to acquire, own, or possess animals in the future.

---

21. G.S. 19A-4. It is possible, though, for a plaintiff to skip the preliminary injunction stage and immediately seek a permanent injunction. It is also possible for a plaintiff to request a permanent injunction after the court has rejected his or her request for a preliminary injunction.

22. Civil cruelty proceedings are held before a judge, not a jury. G.S. 19A-4(a).

23. If the animal is to be returned to the defendant, it is the custodian's responsibility to ensure that it happens. The law states that "[i]f the final judgment entitles the defendant to regain possession of the animal, the custodian shall return the animal, including taking any necessary steps to retrieve the animal from a foster care provider." G.S. 19A-4(c).

24. G.S. 19A-4(d).

25. The order must explain the reasons for the order and describe in detail the act or acts enjoined or restrained. G.S. 1A-1, Rule 65(d).

26. G.S. 19A-4(b). The judge must make this determination based upon a "preponderance of the evidence," which is also referred to as the "greater weight of the evidence." *See, e.g.,* Cincinnati Butchers Supply Co. v. Conoly, 204 N.C. 677, 679, 169 S.E. 415, 416 (1933) (explaining that the two terms are synonymous).

Duty to Exhaust Administrative Remedies

Before initiating the injunction process discussed above, plaintiffs need to exhaust all available administrative remedies. In general, the term *administrative remedies* encompasses remedies that do not involve going to court.[27] Such remedies are outside the traditional judicial system, but when they are available they should be any plaintiff's first step.

In a relatively recent case, for example, an animal welfare organization and an animal welfare advocate brought an action against a local government alleging that its euthanasia procedures constituted cruelty.[28] The court never addressed the merits of the case—whether the government's euthanasia methods did constitute cruelty. Instead, it dismissed the case on the grounds that the plaintiffs should have exhausted their administrative remedies before requesting an injunction. The court explained that, because the county animal control program operated in large part under the authority of local board of health rules that require animals to be euthanized "in a humane manner," the plaintiffs should have first filed an appeal with the board of health for enforcement of the rule.[29]

Another recent case echoed the directive to plaintiffs to exhaust all administrative remedies. In this case, the plaintiffs alleged that a private, nonprofit animal shelter was euthanizing animals in violation of requirements specified in state statute.[30] The court concluded that the plaintiffs should have gone first to the Office of Administrative Hearings—the state office charged with hearing appeals concerning the enforcement of state public health rules. Although the reasoning in this decision is problematic for several reasons, its restatement of the duty to exhaust administrative remedies before filing a complaint in court should be observed.[31]

---

27. *See, e.g.,* G.S. 130A-24 (addressing administrative remedies available in the public health context).

28. Justice for Animals, Inc. v. Robeson County, 164 N.C. App. 366, 595 S.E.2d 773 (2004) ("Specifically, plaintiffs allege that the Robeson County Animal Control Facility injects animals in their hearts without anesthesia resulting in pain, discomfort, and convulsive behavior, and euthanizes cats with a drug not approved for usage on cats."). The plaintiffs also alleged that the county failed to keep adequate records (as required by a local board of health rule) and therefore euthanized animals before the animals' owners had a chance to reclaim them from the shelter. *Id.* at 368, 595 S.E.2d at 775.

29. *Id.* at 371–72, 595 S.E.2d at 777 (citing G.S. 130A-24, which governs appeals related to local board of health rules).

30. Justice for Animals, Inc. v. Lenoir County SPCA, Inc., 168 N.C.App. 298, 300–03, 607 S.E.2d 317, 319–21, *modified and affirmed,* 360 N.C. 48, 619 S.E.2d 494 (2005).

31. The court did not appear to fully understand the state and local public health systems. For example, the decision says that the plaintiffs should have proceeded against "the local board of health in the Office of Administrative Hearings." *Id.* at 304,

Exceptions

The following seven activities are excepted from the civil remedy statute:

- the taking of animals that are under the jurisdiction of the Wildlife Resources Commission;[32]
- activities conducted for purposes of biomedical research or training;
- activities conducted for purposes of production of livestock, poultry, or aquatic species;
- activities conducted for the primary purpose of providing food for human or animal consumption;
- activities conducted for veterinary purposes;
- destruction of any animal for the purposes of protecting the public, other animals, or the public health; and
- activities for sport.[33]

A court may not issue an injunction against a person for participating in any of these activities, as long as the activities are carried out lawfully. But a person could seek an injunction against, for example, a researcher using an animal for biomedical experimentation in a way that is not authorized by law and that causes the animal unjustifiable pain.

Cruelty Investigators

In North Carolina, counties have the option of enlisting private citizens to assist in cruelty investigations. Article 4 of Chapter 19A provides the framework for appointing volunteers as "animal cruelty investigators" and outlines their authority and responsibilities.

Appointment

Boards of county commissioners may appoint one or more persons to serve as animal cruelty investigators. These investigators must serve "without any compensation or other employee benefits," which strongly suggests that they may

---

607 S.E.2d at 321. This statement significantly misstates the relationship between local boards of health (local administrative agencies) and state administrative agencies.

32. The law provides that a "wild bird" exempted from regulation by the Wildlife Resources Commission pursuant to G.S. 113-129(15a) may be the subject of a civil cruelty action. G.S. 19A-1.1(1). The same language is used in the criminal cruelty law and was the subject of extensive litigation. For a detailed discussion of the litigation related to the wild bird provision, see Chapter 1, pp. 7–8.

33. The first six exceptions in the civil statute mirror those in the criminal cruelty statute, G.S. 14-360(c). The seventh exception—lawful activities for sport—is only provided for in the civil remedy.

not be county employees.[34] Even though some jurisdictions have appointed their county animal control officers as cruelty investigators, such appointments should be avoided because they appear to be contrary to the state law establishing the office of cruelty investigator.

The commissioners are allowed to consider candidates nominated by animal welfare organizations, but they may consider other candidates as well.[35] Cruelty investigators are required to take at least six hours a year of continuing education approved by the board of county commissioners. This training must be "designed to give the investigator expertise in the investigation of complaints relating to the care and treatment of animals."[36]

Before making an appointment, the commissioners may choose to enter into an agreement requiring the investigator or an animal welfare organization to assume responsibility for the costs of caring for any animals they seize. Note that this type of agreement is permitted, not required, by state law. The board of commissioners also may agree to reimburse the investigator for necessary and actual expenses related to investigations.[37]

Investigators are appointed for one-year terms and are not limited to any given number of terms. They must take the oath of office and wear badges that (1) are approved by the boards of commissioners and (2) identify them as animal cruelty investigators.[38]

Seizure Authority

A cruelty investigator typically pursues a civil animal cruelty case in the same manner as described in the above section—first seeking a preliminary injunction and then a permanent injunction. Cruelty investigators have, however, one unique authority: they can request, obtain, and execute a seizure order *before* requesting an injunction. To do so, the investigator must first file a sworn complaint with a magistrate. If the magistrate finds "probable cause to believe that the animal is being cruelly treated and that it is necessary for the investigator to

---

34. G.S. 19A-45(a).

35. The law specifically authorizes the board to consider "persons nominated by any society incorporated under North Carolina law for the prevention of cruelty to animals." *Id.*

36. G.S. 19A-49.

37. G.S. 19A-45(d).

38. Chapter 11 of the North Carolina General Statutes governs the administration of oaths to members of the General Assembly and others appointed or elected to public office. It identifies who may administer the oath, defines when affirmation may be substituted for an oath, and prescribes its specific language.

Investigators must supply and pay for their badges "at no cost to the county." G.S. 19A-45(b).

immediately take custody of it," he or she may issue an order authorizing immediate seizure.[39] The order, which is only valid for twenty-four hours, allows the investigator to take custody of and provide suitable care for the animal.

When seizing the animal, the investigator must leave with the owner a copy of the magistrate's order and a written statement describing

- the animal seized
- the place where the animal will be taken
- the reason for taking the animal
- the investigator's intent to file a civil cruelty case

If the investigator does not know who owns the animal, a copy of the above information should be affixed to the premises or vehicle where the animal was found.[40] If anyone is present when the investigator arrives, the investigator must give notice of his or her identity and purpose before entering the premises or vehicle.

When seizing an animal, an investigator may ask to be accompanied by an animal control or law enforcement officer. He or she may *forcibly* enter a building or a vehicle only

- when reasonably certain that the animal is on the premises or in the vehicle,
- when reasonably certain that no people are on the premises or in the vehicle,
- if forcible entry is necessary to seize the animal as authorized by the order,
- when accompanied by a law enforcement officer,
- during daylight hours, and
- when the order is issued by a district court judge (rather than a magistrate).[41]

After seizing the animal, the investigator must return the seizure order to the clerk of court along with a written inventory of the animal or animals seized.[42] The investigator must take the animal to a safe and secure place and provide suitable care for it.

A person who interferes with an animal cruelty investigator in the performance of his or her official duties may be charged with a misdemeanor.[43]

---

39. G.S. 19A-46(a).
40. G.S. 19A-46(c).
41. G.S. 19A–46(b) and (e).
42. G.S. 19A-46(a).
43. G.S. 19A-48 (Class 1 misdemeanor).

Recovering the Custodian's Costs

If a plaintiff assumes custody of an animal during the course of a civil cruelty case, he or she will incur some costs related to its care: food, shelter, and veterinary care. A plaintiff also assumes responsibility for certain court costs and fees associated with bringing the action, although the law provides that such costs need not be paid until the court makes its final decision.[44]

The law provides three mechanisms for recovering some or all of those costs. The first mechanism is fairly simple: if the plaintiff wins a civil animal cruelty case, state law provides that court costs are to be paid by the defendant.[45] Costs typically include the filing fees and other court-related expenses involved in bringing the action.[46] In cruelty cases, however, the judge may also include the costs any food, water, shelter, and care—including veterinary care—the plaintiff provided during the course of the proceeding. If the judge decides to include those expenses as court costs, the defendant will be required to pay them.

The second cost-recovery mechanism is only available to animal cruelty investigators. If an investigator seizes and provides care for an animal during the course of a civil cruelty case, the animal's owner may be held liable for the "necessary expenses" incurred in caring for the animal, including veterinary care.[47] If the animal's owner fails to pay for the care, the investigator may have a lien on the animal. This means that if the owner fails to pay for the care provided to the animal after it was seized, the investigator may be able to sell the animal to recover some or all of the expenses. State law provides a detailed framework for enforcing liens through public or private sale, including deadlines and specific notice requirements.[48]

The third cost-recovery mechanism is the most recent addition to the law and is somewhat more complex than the first two. Basically, it allows certain plaintiffs to limit their own out-of-pocket expenditures by getting money from the defendant up front rather than waiting for the proceeding to conclude. This cost-recovery option is also available to local governments if an animal is seized and sheltered after a person is arrested for (1) criminal cruelty or (2) an

---

44. G.S. 19A-46(d) ("[A]ny person who commences a proceeding under this article [Article 4; animal cruelty investigators] or Article 1 [civil remedy for the protection of animals] shall not be required to pay any court costs or fees prior to a final judicial determination as provided in G.S. 19A-4 [permanent injunction], at which time those costs shall be paid pursuant to the provisions of G.S. 6-18.").

45. G.S. 6-18(5) ("Costs shall be allowed of course to the plaintiff, upon a recovery, in . . . an action brought under Article 1 of Chapter 19A.").

46. *See* G.S. 7A-305 (specifying the court costs that apply in civil actions).

47. G.S. 19A-47.

48. G.S. 44A-4.

attack by a dangerous dog.[49] In the criminal context, the law requires that the defendant be *arrested*, which may present a challenge for jurisdictions that typically address misdemeanor cruelty and dangerous dog cases by issuing criminal summonses rather than arrest warrants. [50] This option is primarily designed to protect the financial interests of local governments and people and organizations that work with local governments. As such, it is only available if the cruelty action is initiated by one of the following:

- a county or municipality,
- a county or municipal official,
- a county-approved animal cruelty investigator, or
- an organization operating a county or municipal shelter under contract.[51]

If one of these four groups files a civil cruelty action and an animal shelter assumes custody of the animal, the shelter operator may petition the court to order the defendant to deposit with the court enough money to cover the "reasonable expenses" of caring for the animal while the litigation is pending. Reasonable expenses include the cost of providing food, water, shelter, and care, including medical care. The initial petition should itemize the costs expected to be incurred for thirty days.

Once such a petition is filed, the court is required to conduct a hearing no earlier than ten business days and no later than fifteen business days after the filing date. The shelter operator must mail notice of the hearing and a copy of the petition to the defendant.[52] At the hearing, the judge should determine how much money is needed to care for the animal for thirty calendar days (not business days). In making this determination, the judge should consider not only the needs of the animal but also the defendant's ability to pay.

At this point in the proceeding, the judge may either

- order the defendant to deposit funds sufficient to care for the animal for thirty days, or
- order the defendant to provide suitable care for the animal at the animal's current location while the litigation is pending—if the judge concludes that the defendant is financially unable to deposit the necessary funds. In conjunction with such an order, an animal control or law enforcement officer must make regular visits to the animal to ensure that it is receiving

49. Article 47 of G.S. 14; G.S. 67-4.3.

50. G.S. 19A-70.

51. G.S. 19A-70(a).

52. If the defendant is in jail, the shelter operator must also provide notice to the custodian of the jail. G.S. 19A-70(b).

proper care.[53] If the officer concludes that the animal is not being cared for appropriately, it may be impounded.

When the judge orders the defendant to deposit funds, the money must be deposited with the clerk of superior court within five days of the initial hearing. Once the funds are posted, the shelter operator is allowed to draw from the funds the actual costs incurred in caring for the animal. If the defendant fails to deposit the funds within that period, the animal is automatically forfeited.

If the case is not resolved within the initial thirty days, the shelter operator may request an extension of the order for additional thirty-day periods until the litigation is resolved. To do so, the operator must file an affidavit with the clerk of superior court stating that to the best of his or her knowledge, the case has not been resolved. This affidavit must be filed at least two business days prior to the expiration of each thirty-day period. Upon receipt of the affidavit, the initial order is automatically renewed for an additional thirty days.

While the litigation is pending, the defendant is required to continue depositing funds within five business days of the end of every thirty-day period, unless the defendant requests a hearing at least five business days before expiration of the period.[54] If the defendant fails to either request a hearing or deposit the funds as required, the animal is automatically forfeited.

Local Laws

Local governments have long had the authority to adopt laws governing the treatment of animals, as discussed in Chapter 1.[55] Violations of local ordinances are often misdemeanors. Some jurisdictions also use such civil law remedies as monetary fines, injunctions, or other forms of equitable relief as alternative mechanisms for enjoining animal cruelty.[56] To avoid challenges to the validity of the ordinance, however, jurisdictions should exercise caution in enforcing any ordinance that duplicates or conflicts with the state law.[57]

---

53. G.S. 19A-70(f).

54. Interestingly, it appears that the defendant's duty to deposit funds every thirty days is independent of the shelter's duty to submit an affidavit requesting extension of the order. This may, however, simply be a drafting error. It would be reasonable for a court to infer that if an order extension is not requested, the original order automatically expires.

55. G.S. 153A-127 (counties); 160A-182 (cities).

56. G.S. 153A-123 (specifying the enforcement mechanisms for local ordinances).

57. G.S. 160A-174(b). See full discussion of local authority in Chapter 1, pp. 17–19.

Conclusion

This chapter, in conjunction with Chapter 1, summarizes and analyzes the many different legal tools available under state law for addressing animal cruelty. In some situations, an injunction may be the most appropriate remedy, while in others a criminal prosecution may prove more effective. Local government officials, animal control officers, and animal cruelty investigators dealing with an instance of animal cruelty will need to understand and consider all the options the law makes available to them.

Relevant Statutes

Article 1 of Chapter 19A
Civil Remedy for the Protection of Animals.

§ 19A-1. Definitions.

The following definitions apply in this Article:

(1) The term "animals" includes every living vertebrate in the classes Amphibia, Reptilia, Aves, and Mammalia except human beings.

(2) The terms "cruelty" and "cruel treatment" include every act, omission, or neglect whereby unjustifiable physical pain, suffering, or death is caused or permitted.

(3) The term "person" has the same meaning as in G.S. 12-3.

§ 19A-1.1. Exemptions.

This Article shall not apply to the following:

(1) The lawful taking of animals under the jurisdiction and regulation of the Wildlife Resources Commission, except that this Article applies to those birds exempted by the Wildlife Resources Commission from its definition of "wild birds" pursuant to G.S. 113-129(15a).

(2) Lawful activities conducted for purposes of biomedical research or training or for purposes of production of livestock, poultry, or aquatic species.

(3) Lawful activities conducted for the primary purpose of providing food for human or animal consumption.

(4) Activities conducted for lawful veterinary purposes.

(5) The lawful destruction of any animal for the purposes of protecting the public, other animals, or the public health.

(6) Lawful activities for sport.

§ 19A-2. Purpose.

It shall be the purpose of this Article to provide a civil remedy for the protection and humane treatment of animals in addition to any criminal remedies that are available and it shall be proper in any action to combine causes of action against one or more defendants for the protection of one or more animals. A real party in interest as plaintiff shall be held to include any person even though the person does not have a possessory or ownership right in an animal; a real party in interest as defendant shall include any person who owns or has possession of an animal.

§ 19A-3. Preliminary injunction; care of animal pending hearing on the merits.

(a) Upon the filing of a verified complaint in the district court in the county in which cruelty to an animal has allegedly occurred, the judge may, as a matter of discretion, issue a preliminary injunction in accordance with the procedures set forth in G.S. 1A-1, Rule 65. Every such preliminary injunction, if the plaintiff so requests, may give the plaintiff the right to provide suitable care for the animal. If it appears on the face of the complaint that the condition giving rise to the cruel treatment of an animal requires the animal to be removed from its owner or other person who possesses it, then it shall be proper for the court in the preliminary injunction to allow the plaintiff to take possession of the animal as custodian.

(b) The plaintiff as custodian may employ a veterinarian to provide necessary medical care for the animal without any additional court order. Prior to taking such action, the plaintiff as custodian shall consult with, or attempt to consult with, the defendant in the action, but the plaintiff as custodian may authorize such care without the defendant's consent. Notwithstanding the provisions of this subsection, the plaintiff as custodian may not have an animal euthanized without written consent of the defendant or a court order that authorizes euthanasia upon the court's finding that the animal is suffering due to terminal illness or terminal injury.

(c) The plaintiff as custodian may place an animal with a foster care provider. The foster care provider shall return the animal to the plaintiff as custodian on demand.

§ 19A-4. Permanent injunction.

(a) In accordance with G.S. 1A-1, Rule 65, a district court judge in the county in which the original action was brought shall determine the merits of the action by trial without a jury, and upon hearing such evidence as may be presented, shall enter orders as the court deems appropriate, including a permanent injunction and dismissal of the action along with dissolution of any preliminary injunction that had been issued.

(b) If the plaintiff prevails, the court in its discretion may include the costs of food, water, shelter, and care, including medical care, provided to the animal, less any amounts deposited by the defendant under G.S. 19A-70, as part of the costs allowed to the plaintiff under G.S. 6-18. In addition, if the court finds by a preponderance of the evidence that even if a permanent injunction were issued there would exist a substantial risk that the animal would be subjected to further cruelty if returned to the possession of the defendant, the court may terminate the defendant's ownership and right of possession of the animal and transfer ownership and right of possession to the plaintiff or other appropriate successor owner. For good cause shown, the court may also enjoin the defendant from acquiring new animals for a specified period of time or limit the number of animals the defendant may own or possess during a specified period of time.

(c) If the final judgment entitles the defendant to regain possession of the animal, the custodian shall return the animal, including taking any necessary steps to retrieve the animal from a foster care provider.

(d) The court shall consider and may provide for custody and care of the animal until the time to appeal expires or all appeals have been exhausted.

Article 4 of Chapter 19
Animal Cruelty Investigators.

§ 19A-45. Appointment of animal cruelty investigators; term of office; removal; badge; oath; bond.

(a) The board of county commissioners is authorized to appoint one or more animal cruelty investigators to serve without any compensation or other employee benefits in his county. In making these appointments, the board may consider persons nominated by any society incorporated under North Carolina law for the prevention of cruelty to animals. Prior to making any such appointment, the board of county commissioners is authorized to enter into an agreement whereby any necessary expenses of caring for seized animals not collectable pursuant to G.S. 19A-47 may be paid by the animal cruelty investigator or by any society incorporated under North Carolina law for the prevention of cruelty to animals that is willing to bear such expense.

(b) Animal cruelty investigators shall serve a one-year term subject to removal for cause by the board of county commissioners. Animal cruelty investigators shall, while in the performance of their official duties, wear in plain view a badge of a design approved by the board identifying them as animal cruelty investigators, and provided at no cost to the county.

(c) Animal cruelty investigators shall take and subscribe the oath of office required of public officials. The oath shall be filed with the clerk of superior court. Animal cruelty investigators shall not be required to post any bond.

(d) Upon approval by the board of county commissioners, the animal cruelty investigator or investigators may be reimbursed for all necessary and actual expenses, to be paid by the county.

§ 19A-46. Powers; magistrate's order; execution of order; petition; notice to owner.

(a) Whenever any animal is being cruelly treated as defined in G.S. 19A-1(2), an animal cruelty investigator may file with a magistrate a sworn complaint requesting an order allowing the investigator to provide suitable care for and take immediate custody of the animal. The magistrate shall issue the order only when he finds probable cause to believe that the animal is being cruelly treated and that it is necessary for the investigator to immediately take custody of it. Any magistrate's order issued under this section shall be valid for only 24 hours after its issuance. After he executes the order, the animal cruelty investigator shall return it with a written inventory of the animals seized to the clerk of court in the county where the order was issued.

(b) The animal cruelty investigator may request a law-enforcement officer or animal control officer to accompany him to help him seize the animal. An investigator may forcibly enter any premises or vehicle when necessary to execute the order only if he reasonably believes that the premises or vehicle is unoccupied by any person and that the animal is on the premises or in the vehicle. Forcible entry shall be used only when the animal cruelty investigator is accompanied by a law- enforcement officer. In any case, he must give notice of his identity and purpose to anyone who may be present before entering said premises. Forcible entry shall only be used during the daylight hours.

(c) When he has taken custody of such an animal, the animal cruelty investigator shall file a complaint pursuant to Article 1 of this Chapter as soon as possible. When he seizes the animal, he shall leave with the owner, if known, or affixed to the premises or vehicle a copy of the magistrate's order and a written notice of a description of the animal, the place where the animal will be taken, the reason for taking the animal, and the investigator's intent to file a complaint in district court requesting custody of the animal pursuant to Article 1 of this Chapter.

(d) Notwithstanding the provisions of G.S. 7A-305(c), any person who commences a proceeding under this Article or Article 1 of this Chapter shall not be required to pay any court costs or fees prior to a final judicial determination as provided in G.S. 19A-4, at which time those costs shall be paid pursuant to the provisions of G.S. 6-18.

(e) Any judicial order authorizing forcible entry shall be issued by a district court judge.

§ 19A-47. Care of seized animals.

The investigator must take any animal he seizes directly to some safe and secure place and provide suitable care for it. The necessary expenses of caring for seized animals, including necessary veterinary care, shall be a charge against the animal's owner and a lien on the animal to be enforced as provided by G.S. 44A-4.

§ 19A-48. Interference unlawful.
It shall be a Class 1 misdemeanor, to interfere with an animal cruelty investigator in the performance of his official duties.

§ 19A-49. Educational requirements.
Each animal cruelty investigator at his own expense must attend annually a course of at least six hours instruction offered by the North Carolina Humane Federation or some other agency. The course shall be designed to give the investigator expertise in the investigation of complaints relating to the care and treatment of animals. Failure to attend a course approved by the board of county commissioners shall be cause for removal from office.

Article 6 of Chapter 19A
Care of Animal Subjected to Illegal Treatment.

§ 19A-70. Care of animal subjected to illegal treatment.

(a) In every arrest under any provision of Article 47 of Chapter 14 of the General Statutes or under G.S. 67-4.3 or upon the commencement of an action under Article 1 of this Chapter by a county or municipality, by a county-approved animal cruelty investigator, by other county or municipal official, or by an organization operating a county or municipal shelter under contract, if an animal shelter takes custody of an animal, the operator of the shelter may file a petition with the court requesting that the defendant be ordered to deposit funds in an amount sufficient to secure payment of all the reasonable expenses expected to be incurred by the animal shelter in caring for and providing for the animal pending the disposition of the litigation. For purposes of this section, "reasonable expenses" includes the cost of providing food, water, shelter, and care, including medical care, for at least 30 days.

(b) Upon receipt of a petition, the court shall set a hearing on the petition to determine the need to care for and provide for the animal pending the disposition of the litigation. The hearing shall be conducted no less than 10 and no more than 15 business days after the petition is filed. The operator of the animal shelter shall mail written notice of the hearing and a copy of the petition to the defendant at the address contained in the criminal charges or the complaint or summons by which a civil action was initiated. If the defendant is in a local detention facility at the time the petition is filed, the operator of the animal shelter shall also provide notice to the custodian of the detention facility.

(c) The court shall set the amount of funds necessary for 30 days' care after taking into consideration all of the facts and circumstances of the case, including the need to care for and provide for the animal pending the disposition of the litigation, the recommendation of the operator of the animal shelter, the estimated cost of caring for and providing for the animal, and the defendant's ability to pay. If the court determines that the defendant is unable to deposit funds, the court may consider issuing an order under subsection (f) of this section.

Any order for funds to be deposited pursuant to this section shall state that if the operator of the animal shelter files an affidavit with the clerk of superior court, at least two business days prior to the expiration of a 30-day period, stating that, to the best of the affiant's knowledge, the case against the defendant has not yet been resolved, the order shall be automatically renewed every 30 days until the case is resolved.

(d) If the court orders that funds be deposited, the amount of funds necessary for 30 days shall be posted with the clerk of superior court. The defendant shall also deposit the same amount with the clerk of superior court every 30 days thereafter until the litigation is resolved, unless the defendant requests a hearing no less than five business days prior to the expiration of a 30-day period. If the defendant fails to deposit the funds within five business days of the initial hearing, or five business days of the expiration of a 30-day period, the animal is forfeited by operation of law. If funds have been deposited in accordance with this section, the operator of the animal shelter may draw from the funds the actual costs incurred in caring for the animal.

In the event of forfeiture, the animal shelter may determine whether the animal is suitable for adoption and whether adoption can be arranged for the animal. The animal may not be adopted by the defendant or by any person residing in the defendant's household. If the adopted animal is a dog used for fighting, the animal shelter shall notify any persons adopting the dog of the liability provisions for owners of dangerous dogs under Article 1A of Chapter 67 of the General Statutes. If no adoption can be arranged after the forfeiture, or the animal is unsuitable for adoption, the shelter shall humanely euthanize the animal.

(e) The deposit of funds shall not prevent the animal shelter from disposing of the animal prior to the expiration of the 30-day period covered by the deposit if the court makes a final determination of the charges or claims against the defendant. Upon determination, the defendant is entitled to a refund for any portion of the deposit not incurred as expenses by the animal shelter. A person who is acquitted of all criminal charges or not found to have committed animal cruelty in a civil action under Article 1 of this Chapter is entitled to a refund of the deposit remaining after any draws from the deposit in accordance with subsection (d) of this section.

(f) Pursuant to subsection (c) of this section, the court may order a defendant to provide necessary food, water, shelter, and care, including any necessary medical care, for any animal that is the basis of the charges or claims against the defendant without the removal of the animal from the existing location and until the charges or claims against the defendant are adjudicated. If the court issues such an order, the court shall provide for an animal control officer or other law enforcement officer to make regular visits to the location to ensure that the animal is receiving necessary food, water, shelter, and care, including any necessary medical care, and to impound the animal if it is not receiving those necessities.

---

§ 153A-127. Abuse of animals.
A county may by ordinance define and prohibit the abuse of animals.

§ 160A-182. Abuse of animals.
A city may by ordinance define and prohibit the abuse of animals.

Chapter 3

Rabies Control

Rabies is a viral infection that may be transmitted to humans through the bite of infected animals such as raccoons, bats, and dogs.[1] If left untreated, the disease is almost always fatal for humans.[2] In North Carolina, public health, animal control, and wildlife management officials work together to enforce state and local laws designed to minimize the spread of rabies and the risk of human exposure.

Like many other states, North Carolina has adopted a series of statutes and regulations governing the control of rabies. These statewide laws not only require the vaccination of cats and dogs but also provide a detailed framework for responding to animal bites and other potential exposures to the rabies virus.

---

1. According to the Centers for Disease Control and Prevention (CDC), the virus can also be contracted when an infected animal's saliva or nervous system tissue comes in direct contact with a person's eyes, nose, mouth, or an open wound. CDC, Questions and Answers about Rabies (July 2, 2004), www.cdc.gov/rabies/qanda/general.html (last visited Jan. 9, 2008) (hereinafter CDC, Rabies Q&A).

2. If a person is bitten by an animal that has or may have rabies, a health care provider will most likely treat the wound and administer a series of vaccines intended to prevent rabies in humans (postexposure antirabies prophylaxis). *See* CDC, "Human Rabies Prevention—United States, 1999: Recommendations of the Advisory Committee on Immunization Practices," *Morbidity and Mortality Weekly Report* 48 (Jan. 8, 1999): 7–13 (hereinafter *MMWR*), www.cdc.gov/rabies/publications/ (last visited Jan. 9, 2008). According to the CDC, the prophylaxis treatment is nearly 100 percent successful. *See* CDC, About Rabies, www.cdc.gov/rabies/about.html (last visited Jan. 9, 2008). Until recently, most people believed that in the absence of prophylaxis treatment the disease would always be fatal to humans. In 2004, however, a girl infected with rabies did not receive the postexposure prophylaxis treatment but still survived. CDC, "Recovery of a Patient from Clinical Rabies—Wisconsin, 2004," *MMWR* 53 (Dec. 24, 2004): 1171–73, www.cdc.gov/mmwr/preview/mmwrhtml/mm5350a1.htm (last visited Jan. 9, 2008); Gretchen Ehlke, "Only Known Unvaccinated Rabies Survivor Thrives," *USA Today*, Dec. 24, 2005.

The relevant statutes can be found in Article 6, Part 6 of Chapter 130A—the public health chapter of the General Statutes.[3] This chapter contains a brief summary of the history of those rabies control laws, provides an overview of their major components, and highlights areas where local government ordinances and board of health rules also play a role.[4]

"Mad" Dogs

The earliest rabies control laws in North Carolina governed the killing of "mad" dogs.[5] Any dog that exhibited symptoms of rabies infection—agitation, loss of appetite, and unusually aggressive behavior—could be considered a "mad" dog.[6]

---

3. The Commission for Health Services has also adopted a few regulations governing rabies control. *See, e.g.*, N.C. Admin. Code tit. 10A, ch. 41G, § .0101–.0103 (hereinafter N.C.A.C.) (addressing rabies vaccinations).

4. Cities and counties have broad authority to adopt ordinances designed to protect the health, safety, and welfare of their citizens. *See* N.C. Gen. Stat. 153A-121 (counties) (hereinafter G.S.); G.S.160A-186 (municipalities). This authority (or "police power") clearly allows local governments to adopt rabies control ordinances that supplement state laws. Boards of health also have authority to adopt local rules in this area. *See* G.S. 130A-39(a) (authorizing boards of health to adopt rules "necessary to protect and promote the public health"). Unlike county ordinances, board of health rules apply to all municipalities within the board's jurisdiction. G.S. 130A-39(c). The scope of board of health rule-making authority, however, has been significantly limited by the courts in the past decade. See below notes 57–59 and related text.

5. The laws date back to the early nineteenth century and have changed very little over time. In 1817, the law read as follows:

> Whereas that most dreadful of all maladies, Hydrophobia, has become much more common than formerly by reason of the negligence of the owners of dogs: For remedy whereof.
>
> 1. Be it enacted. That whenever the owner of any dog shall know, or have good reason to believe, that his or her dog, or any dog belonging to his or her slave, or any other person in his or her employment, has been bitten by a mad dog, and shall neglect or refuse immediately to kill the same, he or she so refusing or neglecting shall pay the sum of twenty-five pounds . . .
> 2. And be it further enacted. That he or she so refusing or neglecting as aforesaid, shall be further liable to pay all damages which may be sustained by any person or persons whatsoever, by the bite of any dog belonging as aforesaid.

A.D. 1817, c. 945.

The fine became fifty dollars in 1837, and the criminal penalty appears to have been added in 1883. R.S. 1837, c. 70 (fine); Code 1883, s. 2499 (misdemeanor).

6. *See, e.g.*, Buck v. Brady, 73 A. 277, 278–79 (Md. Ct. App. 1909) (describing the symptoms of a rabid dog).

North Carolina has two mad dog laws. The first provides that an owner who knows, or has "good reason to believe," that his or her dog has been bitten by a mad dog must immediately kill the dog. Failure to do so may result in both civil and criminal penalties.[7] The second law authorizes any person to kill any mad dog.[8] These two laws predate the modern statutory scheme designed to control rabies and therefore may have been superseded.

Vaccination Requirements

Vaccination of pets is the cornerstone requirement of modern rabies control. Under state law every owner of a dog or cat over four months of age is required to have the animal vaccinated against rabies. According to the office of the state public health veterinarian in the North Carolina Department of Health and Human Services (DHHS), over one million dogs and cats were vaccinated for rabies in 2006.[9]

While state law does not currently require animals other than cats and dogs to be vaccinated, a local government could adopt such a requirement. Some jurisdictions have done so. Charlotte requires vaccination of ferrets, and Buncombe County authorizes the local board of health or health director to order the vaccination of domestic animals other than dogs and cats in the event of a rabies outbreak or epidemic.[10]

Vaccinations may be administered by either licensed veterinarians or by persons who have been appointed "certified rabies vaccinators" by a local health director and trained and certified by the state public health veterinarian.[11] Even though many pet owners take their animals to private veterinarians for rabies vaccinations, local health departments are required to organize (or assist other county departments in organizing) at least one public vaccination clinic per

---

7. G.S. 67-4. The law provides that if the dog that was bitten by a mad dog subsequently bites a person or animal and the owner is sued civilly, the latter must pay the person who was bitten (or the owner of an animal that was bitten) fifty dollars as well as any damages. The owner will also be guilty of a Class 3 misdemeanor. *Id.*

8. G.S. 67-14. The statute also allows the killing of any dog that is killing sheep, cattle, hogs, or poultry. This statute can be traced as far back as 1919. Code, c. 31, art. 3, s. 1682 (1920).

9. Personal communication from Dr. Carl Williams, state public health veterinarian (March 16, 2007). Dr. Williams indicated that the state sold over one million rabies tags in 2006 and that some veterinarians and vaccinators purchase tags from other sources.

10. City of Charlotte Code of Ordinances, § 3-101; Buncombe County Code of Ordinances, § 6-56.

11. G.S. 130A-185; G.S. 130A-186.

year.[12] Boards of county commissioners set the clinic vaccination fee—which is limited by statute to the actual cost of the vaccine, the certificate, and the collar tag, plus an administrative fee of up to four dollars per vaccination. Often the county coordinates these public clinics with the assistance and support of private veterinarians in the community. For example, the county may organize and advertise the clinic, while veterinarians participate at public locations or in their own offices.

Each owner whose pet is vaccinated is given a copy of a vaccination certificate and a rabies tag.[13] At a minimum, the tags must include the year issued, a vaccination number, the words "rabies vaccine" and either "North Carolina" or the initials "N.C."[14]

State law requires dog and cat owners to ensure that their pets wear rabies tags at all times. Local governments may, however, adopt an ordinance exempting cats from the tag requirement.[15] Some local governments have adopted ordinances or board of health rules, for example, that supplement the tag requirement by prohibiting the use of a rabies tag for any animal other than the one that received the vaccination.[16]

People who bring a dog or cat to North Carolina from another state or country are required to either have a certificate from a licensed veterinarian or comply with additional vaccination and confinement requirements.[17] The veterinarian's certificate must demonstrate that the animal

- has been vaccinated in the last year,
- is apparently free from rabies, and
- has not been exposed to rabies.

In the absence of such a certificate, the animal must be securely confined upon entry into the state, vaccinated within one week, and then confined for a further two weeks after vaccination. This law does not apply to animals brought into the state for exhibition purposes as long as they are confined and not permitted to run at-large.

---

12. G.S. 130A-187; G.S. 130A-88.
13. G.S. 130A-189; G.S. 130A-190.
14. G.S. 130A-190.
15. G.S. 130A-190. *See, e.g.*, Cabarrus County Code of Ordinances, § 10-99 (exempting cats from the tag requirement but requiring the owner to maintain the certificates as evidence of vaccination); Guilford County Code of Ordinances, § 5-28(c) (allowing cats to wear an "ear tag" in lieu of the traditional collar tag).
16. *See, e.g.*, Cumberland County Code of Ordinances, § 3-14.
17. G.S. 130A-193.

Exposure and Potential Exposure

To minimize health risks to humans, North Carolina law creates a framework for handling situations in which a person is exposed or potentially exposed to rabies. The components include mandatory reporting, confinement of animals, and, in some cases, destruction of animals. The responsibilities of public health and animal control officials vary depending on the situation and the level of health risk, as outlined below.

If a physician treats a person for an animal bite, and the offending animal is known to be a potential carrier of rabies (such a bat, raccoon, or fox, or another animal that is behaving erratically), the physician has a duty to notify the local health director within twenty-four hours. The report must include the bite victim's name, age and sex.[18]

If a dog or cat bites a person, the victim and the pet's owner (or person possessing or in control of the animal) are both required to notify the local health director immediately.[19] The report must include the names and addresses of the victim and the pet's owner. Note that this law applies to all dogs and cats that bite someone—not just those suspected of having rabies.

After any bite, the offending animal must be confined for ten days. This ten-day period is important because it allows health officials to determine whether the animal was shedding the rabies virus in its saliva at the time of the bite.[20]

The local health director is responsible for designating the place of confinement, which could be a veterinarian's office, a public or private animal shelter, or even the owner's property. Some health directors are comfortable allowing an owner to confine the animal on his or her own property under some circumstances, but many are not. State law leaves this decision entirely up to the health director.[21] Regardless of where the animal is confined, the owner is responsible for any costs related to the confinement.

---

18. G.S. 130A-196.

19. *Id.* If the victim is a minor or is incapacitated, the victim's parent, guardian, or other caregiver may make the report on the victim's behalf.

20. *See North Carolina Manual for Rabies Prevention and Animal Bite Management* (Feb. 2007), 12, www.epi.state.nc.us/epi/vet/pdf/2007NCRabiesManual.pdf (last visited Feb. 28, 2007) (hereinafter *NC Rabies Manual*); *see also* CDC Rabies Q&A (explaining that "an animal may appear healthy, but actually be sick with rabies" because it takes some time for the virus to move from the site of the bite to the brain, where it multiples during the incubation period before beginning to "move from the brain to the salivary glands and saliva [and] the animal begins to show unmistakable signs of rabies").

21. The health director also has the authority to allow "a dog trained and used by a law enforcement agency to be released from confinement to perform official duties upon submission of proof that the dog has been vaccinated for rabies." G.S. 130A-196.

An owner who fails to confine the animal as required by the health director will be guilty of a Class 2 misdemeanor.[22] In addition, the health director has the authority to order seizure and confinement of the animal.

If a dog or cat is exposed to rabies, it may have to be destroyed or quarantined.[23] An animal is considered to have been exposed to rabies if the health director reasonably suspects that the dog or cat was exposed to the saliva or nervous tissue of (1) an animal proven to be rabid or (2) an animal reasonably suspected of having rabies but not available for laboratory diagnosis.[24]

Note that this law provides the local health director with significant discretion in deciding whether an animal has been exposed. If the health director concludes that the dog or cat has been exposed, there are three possible outcomes:

- *Vaccination:* Nothing will happen to the dog or cat if it (1) was vaccinated as required by state law more than three weeks prior to being exposed, and (2) is given a booster dose of rabies vaccine within three days of the exposure.[25]

---

22. *Id.*

23. G.S. 130A-197.

24. According to the Dr. Carl Williams, the state public health veterinarian, a dog or cat will usually be considered to have been exposed to rabies if it

- is bitten by an animal that can be reasonably assumed to have rabies, or
- bites an animal that can be reasonably assumed to have rabies.

A dog or cat would also be considered to be exposed if its mucous membranes (eyes, nose, mouth) came into contact with the saliva or nervous tissue of an animal reasonably suspected of having rabies. If, however, the source animal can be tested for rabies and the results are negative, no exposure would have occurred. E-mail communication from Dr. Carl Williams, Feb. 20, 2007 (on file with author).

25. A state regulation specifies the timetable for required rabies vaccinations. It provides that

> (a) When rabies vaccine is administered by a certified rabies vaccinator to a dog or cat, the dog or cat shall be re-vaccinated annually.
> (b) When rabies vaccine is administered by a licensed veterinarian to a dog or cat, the dog or cat shall be re-vaccinated one year later and every three years thereafter, if a rabies vaccine licensed by the U.S. Department of Agriculture as a three-year vaccine is used. Annual re-vaccination shall be required for all rabies vaccine used other than the U.S. Department of Agriculture three-year vaccine. However, when a local board of health adopts a resolution stating that in order to control rabies and protect the public health annual vaccination is necessary within the area over which they have jurisdiction, then the dog or cat must be vaccinated annually regardless of the type [of] vaccine used, until the resolution is repealed.

10A N.C.A.C. 41G .0101

- *Quarantine:* If the dog or cat has not had the required vaccinations, it may be quarantined at a facility approved by the local health director.[26] The health director must establish the duration of the quarantine (up to six months) and the conditions for the quarantine.
- *Destruction:* If the exposed dog or cat does not meet the vaccination requirements and is not quarantined, it must be humanely destroyed by the owner, an animal control officer, or a peace officer.[27]

It worth noting that, unlike the statute related to confinement of biting animals, this statute does not specifically authorize the health director to allow the animal to be confined on the owner's property. In addition, the quarantine statute refers to a "facility" rather than a "place" designated by the local health director. Based on this word choice and omission of a direct reference to home quarantine, one could infer that the health director is not authorized to allow the animal to be quarantined on the owner's property.

If any animal is suspected of having rabies, the owner or person in possession of the animal must immediately notify the local health director or animal control officer.[28] For example, a person might suspect that his or her dog has rabies because the dog is behaving erratically or exhibiting other symptoms. The health director must then designate a place for the animal to be securely confined for a period of ten days. If the animal dies during that period, its head must be sent to the state's public health laboratory for testing.

With respect to animals other than dogs and cats suspected of rabies exposure, the state public health veterinarian may require the animal in question to be destroyed. Given the limited resources available for testing specimens, the state veterinarian has provided local governments with formal and informal guidance regarding the types of animals that should be destroyed and tested.[29]

---

26. The Centers for Disease Control recommend that the animal be placed in "strict isolation" for six months and vaccinated one month before being released. *See* CDC Rabies Q&A, 2. The six-month period is necessary because the rabies virus may take that long to emerge (i.e., the incubation period). *See NC Rabies Manua*l, 14.

27. While the state's rabies laws do not state that the animal must be killed in a humane manner, they should be read in conjunction with the state's animal cruelty laws, which authorize the lawful destruction of animals to protect the public health but also seek to minimize their "unjustifiable" pain or suffering. *See* G.S. 14-360 (establishing the criminal penalties for cruelty). An animal owner could presumably take the animal to a veterinarian to have it euthanized in a humane manner.

28. G.S. 130A-198. Note that the law requires the notification to go to the *county* animal control officer. But it would certainly be reasonable for an animal owner to notify city animal control instead.

29. *See NC Rabies Manual,* 24. At times, the state public health veterinarian provides additional guidance via e-mail. For example, in May 2006 Dr. Carl Williams transmitted

If an animal is diagnosed by a veterinarian as having rabies, it must be destroyed, and its head must be sent to the state's public health laboratory. Local governments, primarily counties, are responsible for shipping heads to the state.

Rabies Enforcement

Local governments are expected to enforce the statewide rabies laws, but local structures of animal control vary tremendously, and responsibility is often shared between municipalities and counties. In counties, animal control functions may be housed in the health department, the sheriff's office, or a freestanding department. In cities animal control is often under the police department, but it may also be a separate department.

The typical local government has one or more animal control officers or similar officials who impound stray animals without rabies tags and respond to bite reports.[30] If someone is bitten or otherwise potentially exposed to rabies, the local health department often assumes primary responsibility for the communicable disease investigation and response. Some health departments, however, choose to delegate this authority, in whole or in part, to animal control officials. Although the rabies statutes impose many duties on the local health director, that official is allowed to delegate his or her authority to others.[31]

State law requires local government animal control officers to canvass their jurisdictions to find animals not wearing rabies tags.[32] The term *canvass* suggests that officers have a duty to proactively tour the jurisdiction seeking out animals in violation of the law.[33] In practice, however, an animal control officer probably spends more time responding to complaints within the jurisdiction than actually canvassing it.

An officer who finds a dog or cat without a tag has the authority to take action, which may include impounding the animal. If the officer knows who

---

to an animal control listserv an e-mail regarding the testing of opossums (on file with author).

30. Historically, counties employed "dog wardens," who were responsible, at least in part, for rabies control. G.S. 67-31. The dog warden law is still in effect but does not appear to be an integral part of the current animal control system in North Carolina.

31. G.S. 130A-6 ("Whenever authority is granted by this Chapter upon a public official, the authority may be delegated to another person authorized by the public official.").

32. G.S. 130A-192.

33. The dictionary definition of the term *canvass* discusses going through a region to solicit votes, orders, or opinions and also conducting a "thorough examination." *American Heritage Dictionary of the English Language*, 4th ed. (Boston: Houghton Mifflin, 2004), http://dictionary.reference.com/ (last visited Feb. 15, 2007).

owns the animal, he or she is required to notify the owner in writing about the vaccination requirements. The owner must produce a copy of the animal's current vaccination certificate "within three days of the notification."[34] An owner who fails to do so can be charged with a Class 1 misdemeanor.[35] In lieu of criminal prosecution, a local government could also consider seeking an injunction.[36]

If the officer does not know who owns the animal, he or she has the authority to impound it. Note that the officer is not *required* to impound the animal; state law provides local governments with the authority to seize the animal but does not require them to impound all dogs and cats found without rabies tags. An officer who does seize an animal is required to make a reasonable effort to locate the animal's owner.[37]

The local government is required to hold the animal for a time period (at least seventy-two hours) established by the board of county commissioners. Many cities and counties hold animals for longer than seventy-two hours. For example, the city of Garner holds most animals for a minimum of five days, and Durham County holds most animals for ten days.[38] Under federal law, a shelter owned, operated, or under contract to a local government must hold an impounded dog or cat for at least five days before it can sell it to a dealer.[39]

---

34. G.S. 130A-192. The most reasonable interpretation of this language would be that the owner must produce the certificate within three days of *receiving* the notification, rather than within three days after the notification was sent.

35. G.S. 130A-25.

36. G.S. 130A-18.

37. G.S. 130A-192.

38. Garner Code of Ordinances, § 3037; Durham County Code of Ordinances, § 4-40. The Durham County ordinance further provides that if "the animal is not redeemed within twenty-four hours following the last day of confinement, the animal becomes property of the county."

39. 7 U.S.C. § 2158(a). Under the federal law, a "dealer" is defined as

> any person who, in commerce, for compensation or profit, delivers for transportation, or transports, except as a carrier, buys, or sells, or negotiates the purchase or sale of, (1) any dog or other animal whether alive or dead for research, teaching, exhibition, or use as a pet, or (2) any dog for hunting, security, or breeding purposes, except . . .
>
> (i) a retail pet store except such store which sells any animals to a research facility, an exhibitor, or a dealer; or
>
> (ii) any person who does not sell, or negotiate the purchase or sale or any wild animal, dog, or cat and who derives no more than $500 gross income from the sale of other animals during any calendar year.

7 U.S.C. § 2132(f).

If the owner of an impounded animal does not claim the animal within the established time period, the local government may

- allow another person to adopt the animal,
- euthanize the animal, or
- sell the animal to certain research institutions and others registered with the federal government.[40]

Euthanasia

If an animal impounded under the authority of the rabies control law is to be euthanized, state law currently allows local governments to employ any method approved by the American Veterinary Medical Association, the Humane Society of the United States, or the American Humane Association. A chart showing the euthanasia methods approved by these organizations appears in Appendix A. The methods approved by these three organizations vary.[41] Most local government animal shelters in North Carolina use one of two methods:

---

40. State law provides that unclaimed animals may be "sold to institutions within this State registered by the United States Department of Agriculture pursuant to the Federal Animal Welfare Act, as amended." G.S. 130A-192. The Federal Animal Welfare Act (AWA) provides for licensure of dealers and exhibitors and for registration of any research facility, handler, carrier, or exhibitor that is not required to have a license. 7 U.S.C. § 2133 (licenses); 7 U.S.C. § 2136 (registration). Given that the state law refers only to institutions registered pursuant to the AWA, it would be reasonable to interpret North Carolina's law as limiting the sale to research facilities and any institutions that are considered handlers, carriers, or exhibitors. The term *research facility* is defined as

> any school (except an elementary or secondary school), institution, or organization, or person that uses or intends to use live animals in research, tests, or experiments, and that (1) purchases or transports live animals in commerce, or (2) receives funds under a grant, award, loan, or contract from a department, agency, or instrumentality of the United States for the purpose of carrying out research, tests, or experiments.

7 U.S.C. § 2132(e). The secretary of the U.S. Department of Agriculture has the authority to exempt certain persons or entities from the registration requirement in some circumstances.

41. The complete policy statements for all three organizations are available on the Internet. *See* American Humane Association, Animal Welfare Position Statements at www.americanhumane.org/site/DocServer/apsstatements.pdf?docID=101 (last visited Feb. 23, 2007); Humane Society of the United States, The HSUS Statement on Euthanasia Methods for Dogs and Cats (hereinafter HSUS Statement) (Aug. 23, 2005), www.animalsheltering.org/resource_library/policies_and_guidelines/statement_on_euthanasia.html (last visited Feb. 23, 2007); American Veterinary Medical Association, *AVMA Guidelines on Euthanasia* (formerly Report of the AVMA Panel on Euthanasia), www.avma.org/issues/animal_welfare/euthanasia.pdf (last visited Jan. 11, 2008) (hereinafter *AVMA Guidelines*).

(1) sodium pentobarbital injection or (2) inhalation of carbon monoxide gas. Both methods are acceptable under state law.

In response to a 2005 directive from the General Assembly,[42] the Board of Agriculture considered new regulations in February 2008 governing euthanasia of all cats and dogs in animal shelters.[43] The new regulations, which are expected to go into effect sometime in 2008, do not change the list of approved methods of euthanasia; but they do include detailed requirements related to the training and certification of personnel involved in euthanasia, the equipment to be used, and the process to be followed.

The most controversial issue the board faces is whether to allow shelters to continue using carbon monoxide gas. Some oppose the method, arguing that it is inhumane (particularly for very young, old, sick, or pregnant animals) and that exposure to the gas presents too great a health risk for shelter employees.[44] Arguments in favor of carbon monoxide include the relative ease of obtaining the gas as compared to sodium pentobarbital (a controlled substance) and the low cost of obtaining and using the gas.[45] At one point, the regulations proposed prohibiting the use of carbon monoxide after 2012. That provision was removed in the final version considered by the board in February. The draft regulations would require that if carbon monoxide is used (1) only commercially compressed carbon monoxide may be used and (2) the gas must be delivered in a commercially manufactured chamber that allows for the individual separation of animals.[46]

---

42. S.L. 2005-276, sec. 11.5(b); amending G.S. 19A-24.

43. The rules that will amend 02 N.C.A.C. 52J .0203 through .0803 were approved by the Board of Agriculture in February 2008. As of June 2008 they had not been approved by the Rules Review Commission or published in final form in the *North Carolina Register.* For a full explanation of the rule-making process, including the authority of the Rules Review Commission and the General Assembly, see Richard B. Whisnant, *Rule Making in North Carolina* (Chapel Hill: UNC School of Government, 2005), 47–55.

44. *See, e.g.,* HSUS Statement (addressing the risks to both animals and humans); *AVMA Guidelines* at 10 ("Carbon monoxide is extremely hazardous for personnel because it is highly toxic and difficult to detect.").

45. *See* HSUS Statement (recognizing that some shelters do not have ready access to sodium pentobarbital and conditionally accepting carbon monoxide as a" method of euthanasia for some animals when delivered by a commercially manufactured and properly equipped chamber"). *See also* Doug Fakkema, "Comparison of Sodium Pentobarbitol v. Carbon Monoxide as Euthanasia Agents," Killing with Kindness: The Campaign for a Federal Law for the Humane Euthanasia of Companion Animals, www.crean.com/kindness/ebi-vs-co.html (last visited Mar. 7, 2007).

46. G.S. 19A-24.

Rabies in the Community

As we have seen, the basic framework for successful rabies control at the individual animal level lies in the enforcement of preventive vaccination requirements and effective responses to potential rabies exposures. Local governments also have two additional tools they can use to meet the threat of a rabies outbreak in their communities: geographic quarantine and declaration of a rabies emergency.

Geographic Quarantine

A local health director has the authority to order a rabies quarantine for a geographic area if he or she determines that the disease is prevalent enough to endanger the lives of the human population.[47] When an area is under quarantine, dogs and cats must be confined to the owners' premises or a veterinary hospital, kept on a leash, or be under the control and in the sight of a responsible adult. A peace officer or animal control officer who sees a dog or cat running uncontrolled in a quarantined area but cannot catch it is allowed to destroy the animal.

Use of the term *quarantine* in the context of rabies control could be somewhat confusing because the term is used in different ways in other sections of the state's public health laws. Specifically, the term *quarantine authority* defined in Chapter 130A includes orders that

- limit the freedom of movement or action of persons or animals that have been exposed to a communicable disease,
- limit access by any person or animal to an area or facility that may be contaminated with an infectious agent, and
- under certain circumstances, limit the freedom of movement or action of people who have not received immunizations.[48]

The term *quarantine authority* is often used in connection with *isolation authority*.[49] Public health officials have a long history of using these two legal authorities to control the spread of communicable diseases and conditions. The definition above is consistent with the quarantine that a health director might order if a dog or cat has been exposed to rabies or is suspected of having rabies.

However, the definition of quarantine authority above is inconsistent with the geographic quarantine concept in the section of the rabies law that provides for

---

47. G.S. 130A-194; G.S. 130A-195.

48. G.S. 130A-2(7a).

49. *Isolation authority* is available when a person or animal is infected or is reasonably suspected of being infected with rabies. *Quarantine authority* is available when a person or animal has been exposed to or is reasonably suspected of having been exposed to rabies.

restricting the movement and/or action of animals within a certain geographic region. That definition refers specifically to (1) animals that either have been exposed to a communicable disease or have not been immunized and (2) areas and facilities that may be contaminated with an infectious agent. It does *not* address the ability of local health directors to order restrictions for an entire geographic region regardless of exposure to disease or immunization status.

Until this discrepancy in the law is resolved, it would be reasonable for local health directors to assume that they have two separate and distinct quarantine authorities in the limited context of rabies control. The first is the authority to order *all* cats and dogs within a geographic region to be confined or restrained.[50] The second is the more general isolation and quarantine authority available for all communicable diseases, including when a dog or cat is exposed to or suspected of having rabies.[51]

Rabies Emergency

During the early-1990s, a new rabies epidemic began sweeping through North Carolina's wild animal population.[52] In response, the General Assembly enacted a law authorizing local health directors to ask the state health director to declare a rabies emergency in any jurisdiction where rabies is found in a wild animal (other than a bat).[53]

The primary benefit of declaring a rabies emergency in a jurisdiction is that it allows the state health director to ask the Wildlife Resources Commission (WRC) to develop a plan to minimize the threat of rabies exposure to humans

---

50. G.S. 130A-194; G.S. 130A-195.

51. G.S. 130A-197. For more information about this type of quarantine authority, see the discussion above accompanying notes 23–29. For a detailed discussion of the more expansive isolation and quarantine authorities, see Jill Moore, "The North Carolina Public Health System's Isolation and Quarantine Authority," *Health Law Bulletin* No. 84 (July 2006), www.sog.unc.edu/pubs/electronicversions/pdfs/hlb84.pdf (last visited Feb. 20, 2008).

52. According to one report, North Carolina's confirmed cases of rabies almost doubled every year early in the decade. The state had 10 confirmed cases in 1990, 24 cases in 1991, 50 cases in 1992, and 106 cases in 1993. The number peaked in 1997 at 879 confirmed cases. In 2005 there were over 450 confirmed cases. North Carolina Department of Health and Human Services, Division of Public Health, www.epi.state.nc.us/epi/rabies/state.html (last visited Feb. 20, 2007). *See also* Martha Quillin, "Pets Are Vital Link in Battling Rabies," *Raleigh News & Observer* (Mar. 17, 1996), B1 (describing some of the history of the rabies outbreak in the early nineties).

53. G.S. 130A-201. Before declaring an emergency, the state health director must consult with the public health veterinarian (in the N.C. Department of Health and Human Services) and the state agriculture veterinarian (in the N.C. Department of Agriculture and Consumer Services).

and domestic animals from foxes, raccoons, skunks, and/or bobcats.[54] The plan developed by the WRC could, for example, suspend or liberalize hunting restrictions for those animals until the emergency has passed.

Local Rabies Laws

Some local governments have chosen to adopt local ordinances and board of health rules addressing rabies.[55] It may be appropriate to rely on local laws to supplement the state law in some circumstances. For example, a jurisdiction might want to require rabies vaccination of animals other than dogs or cats, establish an extended impoundment period for the shelter, or otherwise address a local concern. Such local laws should not, however, attempt to duplicate, contradict, or change existing state law related to rabies.[56]

Furthermore, a local board of health that wishes to adopt rules governing rabies should ensure that it is acting within the scope of its authority.[57] In a 1996 decision the North Carolina Court of Appeals cautioned boards of health not to consider factors other than health when establishing local rules.[58] Note that an elected body, such as a board of county commissioners or a city council, is not subject to that inherent limitation on its authority and therefore may go beyond human health concerns when adopting ordinances.

The North Carolina Supreme Court has also recently addressed limitations on local board of health rule-making authority. The court indicated that a local board of health that wishes to regulate a field already heavily regulated by the state must base its additional regulation on a particular local health need. In other words, the board must provide "a rationale or basis for making the [local rules] more rigorous than those applicable to and followed by the rest of the state."[59]

---

54. The Wildlife Resources Commission has the authority to develop such plans pursuant to G.S. 113-291.2(a1).

55. Local governments do not have specific authority to regulate in this area, though elected bodies may rely on their general ordinance-making powers and boards of health upon their general rule-making authority. G.S. 130A-39 (boards of health); 153A-121 (county authority); 160A-174 (city authority).

56. *See* G.S. 160-174(b) (identifying the preemption principles applicable to city ordinances); State v. Tenore, 280 N.C. 238, 247, 185 S.E.2d 644, 650 (1972) (extending those preemption principles to county ordinances).

57. For more information regarding the rule-making authority of boards of health, see Aimee Wall, "The Rulemaking Authority of North Carolina Local Boards of Health," *Health Law Bulletin* No. 81 (Nov. 2003), www.sog.unc.edu/pubs/electronicversions/pdfs/hlb81.pdf.

58. City of Roanoke Rapids v. Peedin, 124 N.C. App 578, 589, 478 S.E.2d 528, 535 (1996).

59. Craig v. County of Chatham, 356 N.C. 40, 565 S.E.2d 172 (2002).

Relevant Statutes

G.S. 67–4. Failing to kill mad dog.
If the owner of any dog shall know, or have good reason to believe, that his dog, or any dog belonging to any person under his control, has been bitten by a mad dog, and shall neglect or refuse immediately to kill the same, he shall forfeit and pay the sum of fifty dollars ($50.00) to him who will sue therefor; and the offender shall be liable to pay all damages which may be sustained by anyone, in his property or person, by the bite of any such dog, and shall be guilty of a Class 3 misdemeanor.

§ 67-14. Mad dogs, dogs killing sheep, etc., may be killed.
Any person may kill any mad dog, and also any dog if he is killing sheep, cattle, hogs, goats, or poultry.

Part 6 of Chapter 130A
Rabies.

§ 130A-184. Definitions.
The following definitions shall apply throughout this Part:

(1) "Animal Control Officer" means a city or county employee designated as dog warden, animal control officer, animal control official or other designations that may be used whose responsibility includes animal control.
(2) "Cat" means a domestic feline.
(3) "Certified rabies vaccinator" means a person appointed and certified to administer rabies vaccine to animals in accordance with this Part.
(4) "Dog" means a domestic canine.
(5) "Rabies vaccine" means an animal rabies vaccine licensed by the United States Department of Agriculture and approved for use in this State by the Commission.
(6) "State Public Health Veterinarian" means a person appointed by the Secretary to direct the State public health veterinary program.
(7) "Vaccination" means the administration of rabies vaccine by a licensed veterinarian or by a certified rabies vaccinator.

§ 130A-185. Vaccination of all dogs and cats.
(a) The owner of every dog and cat over four months of age shall have the animal vaccinated against rabies. The time or times of vaccination shall be established by the Commission. Rabies vaccine shall be administered only by a licensed veterinarian or by a certified rabies vaccinator.

(b) Only animal rabies vaccine licensed by the United States Department of Agriculture and approved by the Commission shall be used on animals in this State.

§ 130A-186. Appointment and certification of certified rabies vaccinator.
In those counties where licensed veterinarians are not available to participate in all scheduled county rabies control clinics, the local health director shall appoint one or more persons for the purpose of administering rabies vaccine to animals in that county. Whether or not licensed veterinarians are available, the local health director may appoint one or more persons for the purpose of administering rabies vaccine to animals in their county and these persons will make themselves available to participate in the county rabies control program. The State Public Health Veterinarian shall provide at least four hours of training to those persons appointed by the local health director to administer rabies vaccine. Upon satisfactory completion of the training, the State Public Health Veterinarian shall certify in writing that the appointee has demonstrated a knowledge and procedure acceptable for the administration of rabies vaccine to animals. A certified rabies vaccinator shall be authorized to administer rabies vaccine to animals in the county until the appointment by the local health director has been terminated.

§ 130A-187. County rabies vaccination clinics.
The local health director shall organize or assist other county departments to organize at least one countywide rabies vaccination clinic per year for the purpose of vaccinating dogs and cats. Public notice of the time and place of rabies vaccination clinics shall be published in a newspaper having general circulation within the area.

§ 130A-188. Fee for vaccination at county rabies vaccination clinics.
The county board of commissioners is authorized to establish a fee to be charged at the county rabies vaccination clinics. The fee shall include an administrative charge not to exceed four dollars ($4.00) per vaccination, and a charge for the actual cost of the vaccine, the vaccination certificate, and the rabies vaccination tag.

§ 130A-189. Rabies vaccination certificates.
A licensed veterinarian or a certified rabies vaccinator who administers rabies vaccine to a dog or cat shall complete a three-copy rabies vaccination certificate. The original rabies vaccination certificate shall be given to the owner of each dog or cat that receives rabies vaccine. One copy of the rabies vaccination

certificate shall be retained by the licensed veterinarian or the certified rabies vaccinator. The other copy shall be given to the county agency responsible for animal control, provided the information given to the county agency shall not be used for commercial purposes.

§ 130A-190. Rabies vaccination tags.

(a) A licensed veterinarian or a certified rabies vaccinator who administers rabies vaccine to a dog or cat shall issue a rabies vaccination tag to the owner of the animal. The rabies vaccination tag shall show the year issued, a vaccination number, the words "North Carolina" or the initials "N.C." and the words "rabies vaccine." Dogs and cats shall wear rabies vaccination tags at all times. However, cats may be exempted from wearing the tags by local ordinance.

(b) Rabies vaccination tags, links and rivets may be obtained from the Department. The Secretary is authorized to establish by rule a fee for the rabies tags, links and rivets. Except as otherwise authorized in this section, the fee shall not exceed the actual cost of the rabies tags, links and rivets, plus transportation costs. The Secretary may increase the fee beyond the actual cost plus transportation, by an amount not to exceed five cents ($.05) per tag, to fund rabies education and prevention programs.

(c) The Department shall make available a special edition rabies tag to be known as the "I Care" tag. This tag shall be different in shape from the standard tag and shall carry the inscription "I Care" in addition to the information required by subsection (a) of this section. The Secretary is authorized to establish a fee for the "I Care" rabies tag equal to the amount set forth in subsection (b) of this section plus an additional fifty cents ($.50). The additional fifty cents ($.50) shall be credited to the Spay/Neuter Account established in G.S. 19A-62.

§ 130A-191. Possession and distribution of rabies vaccine.

It shall be unlawful for persons other than licensed veterinarians, certified rabies vaccinators and persons engaged in the distribution of rabies vaccine to possess rabies vaccine. Persons engaged in the distribution of vaccines may distribute, sell and offer to sell rabies vaccine only to licensed veterinarians and certified rabies vaccinators.

§ 130A-192. Dogs and cats not wearing required rabies vaccination tags.

The Animal Control Officer shall canvass the county to determine if there are any dogs or cats not wearing the required rabies vaccination tag. If a dog or cat is found not wearing the required tag, the Animal Control Officer shall check to see if the owner's identification can be found on the animal. If the animal is wearing an owner identification tag, or if the Animal Control Officer otherwise knows who the owner is, the Animal Control Officer shall notify the owner in writing to have the animal vaccinated against rabies and to produce the required rabies vaccination certificate to the Animal Control Officer within three days of the notification. If the animal is not wearing an owner identification tag and the Animal Control Officer does not otherwise know who the owner is, the Animal Control Officer may impound the animal. The duration of the impoundment of these animals shall be established by the county board of commissioners, but the duration shall not be less than 72 hours. During the impoundment period, the Animal Control Officer shall make a reasonable effort to locate the owner of the animal. If the animal is not reclaimed by its owner during the impoundment period, the animal shall be disposed of in one of the following manners: returned to the owner; adopted as a pet by a new owner; sold to institutions within this State registered by the United States Department of Agriculture pursuant to the Federal Animal Welfare Act, as amended; or put to death by a procedure approved by the American Veterinary Medical Association, the Humane Society of the United States or of the American Humane Association. The Animal Control Officer shall maintain a record of all animals impounded under this section which shall include the date of impoundment, the length of impoundment, the method of disposal of the animal and the name of the person or institution to whom any animal has been released.

§ 130A-193. Vaccination and confinement of dogs and cats brought into this State.

(a) A dog or cat brought into this State shall immediately be securely confined and shall be vaccinated against rabies within one week after entry. The animal shall remain confined for two weeks after vaccination.

(b) The provisions of subsection (a) shall not apply to:

(1) A dog or cat brought into this State for exhibition purposes if the animal is confined and not permitted to run at large; or

(2) dog or cat brought into this State accompanied by a certificate issued by a licensed veterinarian showing that the dog or cat is apparently free from and has not been exposed to rabies and that the dog or cat has received rabies vaccine within the past year.

§ 130A-194. Quarantine of districts infected with rabies.

An area may be declared under quarantine against rabies by the local health director when the disease exists to the extent that the lives of persons are endangered. When quarantine is declared, each dog and cat in the area shall be confined on the premises of the owner or in a veterinary hospital. However, dogs or cats on a leash or under the control and in the sight of a responsible adult may be permitted to leave the premises of the owner or the veterinary hospital.

§ 130A-195. Destroying stray dogs and cats in quarantine districts.
When quarantine has been declared and dogs and cats continue to run uncontrolled in the area, any peace officer or Animal Control Officer shall have the right, after reasonable effort has been made to apprehend the animals, to destroy the uncontrolled dogs and cats and properly dispose of their bodies.

§ 130A-196. Confinement of all biting dogs and cats; notice to local health director; reports by physicians; certain dogs exempt.
When a person has been bitten by a dog or cat, the person or parent, guardian or person standing in loco parentis of the person, and the person owning the animal or in control or possession of the animal shall notify the local health director immediately and give the name and address of the person bitten and the owner of the animal. All dogs and cats that bite a person shall be immediately confined for 10 days in a place designated by the local health director. However, the local health director may authorize a dog trained and used by a law enforcement agency to be released from confinement to perform official duties upon submission of proof that the dog has been vaccinated for rabies in compliance with this Part. After reviewing the circumstances of the particular case, the local health director may allow the owner to confine the animal on the owner's property. An owner who fails to confine his animal in accordance with the instructions of the local health director shall be guilty of a Class 2 misdemeanor. If the owner or the person who controls or possesses a dog or cat that has bitten a person refuses to confine the animal as required by this section, the local health director may order seizure of the animal and its confinement for 10 days at the expense of the owner. A physician who attends a person bitten by an animal known to be a potential carrier of rabies shall report within 24 hours to the local health director the name, age and sex of that person.

§ 130A-197. Infected dogs and cats to be destroyed; protection of vaccinated dogs and cats.
When the local health director reasonably suspects that a dog or cat has been exposed to the saliva or nervous tissue of a proven rabid animal or animal reasonably suspected of having rabies that is not available for laboratory diagnosis, the dog or cat shall be considered to have been exposed to rabies. A dog or cat exposed to rabies shall be destroyed immediately by its owner, the county Animal Control Officer or a peace officer unless the dog or cat has been vaccinated

against rabies in accordance with this Part and the rules of the Commission more than three weeks prior to being exposed, and is given a booster dose of rabies vaccine within three days of the exposure. As an alternative to destruction, the dog or cat may be quarantined at a facility approved by the local health director for a period up to six months, and under reasonable conditions imposed by the local health director.

§ 130A-198. Confinement.

A person who owns or has possession of an animal which is suspected of having rabies shall immediately notify the local health director or county Animal Control Officer and shall securely confine the animal in a place designated by the local health director. Dogs and cats shall be confined for a period of 10 days. Other animals may be destroyed at the discretion of the State Public Health Veterinarian.

§ 130A-199. Rabid animals to be destroyed; heads to be sent to State Laboratory of Public Health.

An animal diagnosed as having rabies by a licensed veterinarian shall be destroyed and its head sent to the State Laboratory of Public Health. The heads of all dogs and cats that die during the 10-day confinement period required by G.S. 130A-196, shall be immediately sent to the State Laboratory of Public Health for rabies diagnosis.

§ 130A-201. Rabies emergency.

A local health director in whose county or district rabies is found in the wild animal population as evidenced by a positive diagnosis of rabies in the past year in any wild animal, except a bat, may petition the State Health Director to declare a rabies emergency in the county or district. In determining whether a rabies emergency exists, the State Health Director shall consult with the Public Health Veterinarian and the State Agriculture Veterinarian and may consult with any other source of veterinary expertise the State Health Director deems advisable. Upon finding that a rabies emergency exists in a county or district, the State Health Director shall petition the Executive Director of the Wildlife Resources Commission to develop a plan pursuant to G.S. 113-291.2(a1) to reduce the threat of rabies exposure to humans and domestic animals by foxes, raccoons, skunks, or bobcats in the county or district. Upon determination by the State Health Director that the rabies emergency no longer exists for a county or district, the State Health Director shall immediately notify the Executive Director of the Wildlife Resources Commission.

Chapter 4

Dangerous Dogs

Every North Carolina local government has probably faced the problem of a dog that is threatening or dangerous to persons or other animals. State law provides a ready framework for handling these situations;[1] it defines the term *dangerous dog* and imposes certain restrictions and obligations on the owners of such dogs. Many local governments as well have developed systems for addressing what can become a complex problem. This chapter reviews the state's framework for dealing with dangerous dogs and discusses some of the different approaches adopted in local ordinances across the state. It also briefly discusses the authority of local governments to adopt breed-specific laws, such as ordinances banning private ownership of pit bulls and other breeds. The chapter does not address the law governing civil claims for money damages against owners whose dogs may have harmed a person or damaged property.

State Law

Definition

Before describing how the state law addresses dangerous dogs, it is important to understand how certain terms are defined and used in the law. The statutory definition of the term *dangerous dog* is rather confusing. A dangerous dog is one that

- is owned or harbored primarily or in part for the purpose of dogfighting,
- is trained for dogfighting,
- has, without provocation, killed or inflicted severe injury on a person, or
- is determined to be potentially dangerous by a person or board authorized by a local government to make such judgments.[2]

---

1. Article 1A of N.C. Gen. Stat. ch. 67 (hereinafter G.S.).
2. G.S. 67-4.1(a)(1).

The first two categories are relatively clear. If a dog is owned, harbored, or trained for dogfighting, it is automatically considered a dangerous dog under the law. In the absence of a criminal conviction for dogfighting, however, it may be difficult to prove that a dog falls within one of these categories. It also may be difficult to prove, at times, who owns, harbors, or trains a particular dog. The law defines the term *owner* as "any person or legal entity that has a possessory property right in a dog."[3] In *Lee v. Rice*, the Court of Appeals concluded that a person who simply owned property where a dangerous dog was housed could not be held civilly liable as an "owner or keeper" without more evidence that the property owner exercised some control over or management of the dog.[4] The court would most likely apply the same analysis to the term as used throughout the dangerous dog law—requiring evidence of some level of responsibility for the dog's care before recognizing a person as an owner.

The third category includes dogs that have killed or inflicted severe injury on a person. According to the statute, a *severe injury* is a physical injury that either (1) resulted in broken bones or disfiguring lacerations or (2) required cosmetic surgery or hospitalization. This third category also requires that the killing or injury be unprovoked. Inclusion of the word *unprovoked* is largely redundant, because most types of provoked attacks are already subject to an exception. The law specifically does not apply when the injury inflicted by the dog was sustained by someone who was "committing a willful trespass or other tort or crime [or] was tormenting, abusing, or assaulting the dog" at the time of the injury, or had done so in the past.[5]

The fourth category is perhaps the most confusing section of the state law. The first three categories described above provide that any dog that meets certain criteria is automatically considered a dangerous dog. The fourth category is different in that it requires a local government official or board to make an official determination that the dog meets the definition of "potentially dangerous dog" set out in the statute. If the dog meets the definition, it will be classified as a potentially dangerous dog, which is one type of dangerous dog. In other words, if a dog is found to be potentially dangerous, it will be treated as dangerous for the purposes of enforcing state law.

---

3. G.S. 67-4.1(a)(3).

4. 154 N.C. App. 471, 475–76, 572 S.E.2d 219, 222–23 (2002).

5. G.S. 67-4.1(b)(4). The drafting of the exception is awkward in that the it is unclear whether the clause "at the time of the injury" modifies all of the provocative behaviors or only the trespass-related behavior. For the purposes of this summary, it seems reasonable to conclude that the clause does *not* modify the language referring to a person who "had tormented, abused, or assaulted" the dog because the language is clearly in the past tense.

Three types of behavior trigger an animal's designation as a potentially dangerous dog.

- Inflicting a bite on a *person* that results in broken bones or disfiguring lacerations or requires cosmetic surgery or hospitalization. Note that this language is almost identical to the definition of *severe injury* that is incorporated by reference in the definition of dangerous dog. Therefore, it appears that a dog that inflicts such a physical injury may automatically be considered a dangerous dog, *or* it may be declared a potentially dangerous dog.
- Killing or inflicting severe injury on a domestic animal when not on the owner's real property. The definition of severe injury discussed above applies in this context as well (i.e., broken bones, disfiguring lacerations, cosmetic surgery, or hospitalization). A key component of this type of behavior is that it occurs when the dog is not on the owner's property.
- Approaching a person when not on the owner's property in a vicious or terrorizing manner in an apparent attitude of attack.[6] In *Caswell County v. Hanks,* the court of appeals upheld the constitutionality of this part of the definition of potentially dangerous dog. The court explained that the law was not unconstitutionally vague because it "provides sufficient notice for defendants and others to determine what conduct is proscribed."[7]

As mentioned above, a dog will be considered potentially dangerous if a person or board makes such a determination on behalf of a local government. The law neither prescribes a process for making such determinations nor identifies a particular type of person or board to be charged with this responsibility. In some jurisdictions, an individual animal control officer or supervisor may make the determination; in others, a board established for this purpose does so. If a dog is found to be potentially dangerous, the local government must notify the owner in writing of (1) the determination and (2) the reasons for the determination. While not specifically required in the statute, the notice should also explain the process for appealing the determination.

Local governments are required to designate a board to hear appeals from determinations that a dog is potentially dangerous. The law does not dictate the number of people that must serve on the appeals board or the type of person or professional that must be represented. The law does, however, exclude

---

6. Legislation introduced in 2007 proposed to eliminate this type of threatening behavior from the scope of the state's dangerous dog law. S 92. The bill passed the Senate and could be considered by the House in the 2008 short session.

7. *Hanks*, 120 N.C. App. 489, 493, 462 S.E.2d 841, 844 (1995).

individuals who were involved in the initial determination.[8] All other decisions related to the size and composition of the appeals board are left to the local government.

To appeal a determination that a dog is potentially dangerous, the owner must file written objections with the appeals board "within three days." Because this "three day" language is rather vague, it would be prudent for the local government to provide more specific guidance to the owner at the time of the determination. For example, the local government should decide whether

- the three-day period starts on the day of the determination or several days after the determination (particularly if the determination is mailed to the owner),
- the three days are calendar days or working days, and
- the written objections must be postmarked or delivered to the appeals board by the third day.

This level of specificity in the written notice should help the owner understand more clearly the appeals process.

Once the owner has filed an objection, the appeals process begins and the board must schedule a hearing within ten days. If the appeals board upholds the initial determination, the owner has ten days to file a notice of appeal and petition for review in superior court. A superior court judge will not review the evidence and information collected by the appeals board but will conduct a de novo review, hearing "the case on its merits from beginning to end as if no hearing had been held by the Board and without any presumption in favor of the Board's decision."[9] Decisions made at the superior court level may be further appealed to the court of appeals and the supreme court.

The three critical definitions—*dangerous dog, potentially dangerous dog,* and *severe injury*—overlap in a way that sometimes makes it difficult to understand precisely how the law should be applied. In summary, the following types of dogs are considered dangerous dogs under the state's statutory scheme:

- a dog owned or harbored primarily or in part for the purpose of dogfighting,

---

8. The statute directs the local government to designate "a separate Board to hear any appeal." G.S. 67-4.1(c). It is clear from this language that the membership of an appeals board should be different from the designation board. As a result, it is possible to infer that independence of the appeals board is expected. Therefore, in jurisdictions that allow one individual (rather than a board) to make the initial determination that a dog is potentially dangerous, it seems appropriate to exclude that person from the appeals board.

9. *Hanks*, 120 N.C. App. at 491, 462 S.E.2d at 843.

- a dog trained for dogfighting,
- a dog that without provocation has killed or inflicted severe injury on a person,
- a dog a local government representative or board decides has inflicted severe injury on a person,
- a dog a local government representative or board decides has killed or inflicted severe injury upon a domestic animal when not on the owner's property, or
- a dog a local government representative or board decides approached a person when not on the owner's property in a vicious or terrorizing manner in an apparent attitude of attack.

There are a few categories of dogs that will not be considered dangerous dogs even if they satisfy one of the criteria identified above.[10] As discussed earlier, dogs that are provoked by certain behavior may not be classified as dangerous. Specifically, a dog will not be considered dangerous if it has injured a person who

- at the time of the injury, was committing a willful trespass or other tort,
- at the time of the injury, was tormenting, abusing, or assaulting the dog,
- had tormented, abused, or assaulted the dog in the past, or
- was committing or attempting to commit a crime.

The law will also not recognize a dog as dangerous when it is used by a law enforcement officer to carry out official duties or is used in a lawful hunt. Finally, the state law will not apply to a dog that acted when it (1) was on its owner's property or under its owner's control, (2) was working as a hunting, herding, or predator-control dog, and (3) damaged or injured a domestic animal that is of the species or type of domestic animal appropriate to the work of the dog (i.e., activities related to hunting, herding, and predator control).

Consequences for Owners of Dangerous Dogs

Under state law, an owner of a dangerous dog faces several important restrictions as well as potential civil and criminal consequences. First, the dog must not be left alone on the owner's property unless it is indoors or in a securely enclosed and locked pen or other structure. Second, the dog must not be allowed off the owner's property unless it is leashed (or otherwise restrained) and muzzled. Third, if the owner sells or gives the dog to someone else, he or she must notify the local government in writing about the change and notify the person who is taking possession of the dog, also in writing, about the dog's

10. G.S. 67-4.1(b).

dangerous behavior and, if applicable, the local government's determination that the dog is potentially dangerous. It is a Class 3 misdemeanor to fail to comply with these restrictions and notification requirements.[11]

If a dog that has been determined to be dangerous under the law subsequently attacks someone and causes physical injuries that require medical care costing more than one hundred dollars, its owner may be charged with a Class 1 misdemeanor.

In addition to the potential criminal liability described above, an owner of a dangerous dog may be sued in civil court for money damages for harm to persons or property caused by the dog. In such cases, the owner is subject to "strict liability," which means that the person seeking damages from the owner is not required to show that the owner was negligent in some manner, such as allowing the dog to escape.[12] The court should award damages simply based on the fact that the dog is a dangerous dog under the state law and injured a person or property.

Before the state established the legal framework governing dangerous dogs, it relied in large part on authority granted to local public health directors to declare any animal "vicious and a menace to the public health." This authority is still available when an animal has made an unprovoked attacked on a person and caused bodily harm.[13] Once an animal has been declared vicious, it must be confined to its owner's property except when (1) accompanied by a responsible adult and (2) restrained on a leash. Though local health directors still have this authority, they rarely use it to address problems with dogs. It is important to note that neither the dangerous dog law nor the authority to declare an animal vicious authorizes a local government to euthanize a dangerous dog.

Local Ordinances

Many local governments have adopted dangerous dog ordinances over the years. Cities and counties may rely on two sources of statutory authority for these laws. First, they may exercise their police powers to adopt ordinances to "define, prohibit, regulate, or abate acts, omissions, or conditions detrimental to the health, safety, or welfare of [their] citizens and the peace and dignity" of the city or county and to "define and abate nuisances."[14] Second, local govern-

---

11. G.S. 67-4.2.

12. G.S. 67-4.4.

13. G.S. 130A-200. The statute does not use the term *unprovoked* but rather provides that the animal must not have been "teased, molested, provoked, beaten, tortured, or otherwise harmed."

14. G.S. 153A-121 (counties); G.S. 160A-174 (cities).

ments have specific authority to "regulate, restrict, or prohibit the possession or harboring . . . of animals which are dangerous to persons or property."[15] A number of local governments have ordinances that were already in place when the state's statutory framework was enacted in 1989, and some ordinances have been adopted or amended since that time. State law does not override the local ordinances in this field but, rather, provides that the state statutes must not be "construed to prevent a city or county from adopting or enforcing its own program for control of dangerous dogs."[16]

Some cities and counties have adopted detailed ordinances that are quite different from the state statutory scheme. For example, Boiling Springs Lake requires owners of dangerous dogs to register their animals with city animal control officials, obtain a permit, and comply with several additional restrictions and limitations on the dog's freedom.[17] Some jurisdictions supplement or modify the state law by adopting different definitions, imposing additional restrictions, or outlining administrative procedures that apply to dangerous dog determinations. For example:

- Catawba County's ordinance specifically states that "voice command" does not constitute adequate restraint when a dangerous dog is off the owner's property.[18]
- Garner requires owners of dangerous dogs to notify animal control officials if the dog attacks a person or animal, causes property damage, or escapes from confinement or restraint.[19]
- Durham requires that "dogs declared dangerous or potentially dangerous pursuant to this article must be permanently identified by a microchip implanted under the dog's skin within 30 days following the final determination of dangerousness."[20]
- Fayetteville's ordinance outlines procedures and guidelines governing hearings conducted by the appeals board.[21]
- Guilford County's ordinance includes the following design specifications for enclosures housing dangerous dogs:

  The structure must be a minimum size of fifteen feet by six feet by six feet (15' X 6' X 6') with a floor consisting of a concrete pad at least four inches thick. If more than one animal is to be kept in the enclosure,

---

15. G.S. 153A-131 (counties); G.S 160A-187 (cities).
16. G.S. 67-4.5.
17. Boiling Springs Lake Code of Ordinances, §§ 3-101 to 3-107.
18. Catawba County Code of Ordinances, § 6-132(2).
19. Garner Code of Ordinances, § 3-20.
20. Durham Code of Ordinances, § 4-195.
21. Fayetteville Code of Ordinances, § 6-99.

the floor area must provide at least 45 square feet for each animal. The walls and roof of the structure must be constructed of welded chain link of a minimum thickness of 12 gauge, supported by galvanized steel poles at least 2 1/2 inches in diameter. The vertical support poles must be sunk in concrete-filled holes at least 18 inches deep and at least eight inches in diameter. The chainlink fencing must be anchored to the concrete pad with galvanized steel anchors placed at intervals of no more than 12 inches along the perimeter of the pad. The entire structure must be freestanding and not be attached or anchored to any existing fence, building, or structure. The structure must be secured by a child-resistant lock.[22]

- Laurinburg requires owners of dangerous dogs to permit animal control officials to inspect the owner's premises as necessary to ensure compliance with the state law and local ordinance.[23]

A few jurisdictions have adopted ordinances that simply reiterate the state's dangerous dog law. This approach is not recommended because the "elements of an offense defined by a city ordinance [may not be] identical to the elements of an offense defined by State or federal law."[24]

Several local governments across the state have considered, but rejected, the idea of adopting ordinances restricting or prohibiting private citizens from owning specific breeds of dogs, such as pit bulls. The discussion typically begins after someone in the city or county is attacked by a particular breed of dog. A local government considering such ordinances must decide whether (1) breed-specific legislation is the best policy tool available for addressing the jurisdiction's concerns and (2) whether it has the authority to adopt an ordinance restricting that type of animal. With respect to the first issue, advocates of breed-specific legislation often cite dog-bite statistics related to particular breeds,[25] while those opposed argue that more comprehensive legislation addressing dangerous dogs, animal bites, and owner behavior will have a greater impact.[26]

---

22. Guilford County Code of Ordinances, § 5-9.

23. Laurinburg Code of Ordinances, § 4-15(d)(4)

24. *See* G.S. 160-174(b) (identifying the preemption principles applicable to city ordinances); State v. Tenore, 280 N.C. 238, 247, 185 S.E.2d 644, 650 (1972) (extending those same preemption principles to county ordinances).

25. *See, e.g.,* Jeffrey J. Sacks et al., "Breeds of dogs involved in fatal human attacks in the United States between 1979 and 1998," *Journal of the American Veterinary Medical Association* 217 (Sept. 15, 2000): 836–40.

26. *See, e.g.,* Devin Burstein, "Breed Specific Legislation: Unfair Prejudice and Ineffective Policy," *Animal Law* 10 (2004): 313; Linda S. Weiss, "Breed-Specific Legislation in

With respect to second issue—the scope of the government's authority—North Carolina local governments probably do have the authority to enact breed-specific ordinances.[27] As discussed above, cities and counties have broad authority to regulate dogs within their jurisdictions through both their general police powers and the authority to regulate dangerous animals. This authority would probably extend to regulation of particular breeds if the jurisdiction has a rational reason for believing that the breed is dangerous and that regulation is needed to protect the public. Before moving forward with such an ordinance, though, a local government should review breed-specific laws in other states and carefully craft the ordinance to avoid potential constitutional defects.[28]

---

the United States," Animal Legal and Historical Web Center (2001), www.animallaw.info/articles/aruslweiss2001.htm (last visited Oct. 12, 2007); Humane Society of the United States, Statement on Dangerous Dogs and Breed-Specific Legislation, www.hsus.org/pets/issues_affecting_our_pets/dangerous_dogs.html (last visited Oct. 12, 2007).

27. Jeanette Cox, "Ordinances Targeting Pit Bull Dogs Must Be Drafted Carefully," *Local Government Law Bulletin* No. 106 (December 2004), www.sog.unc.edu/pubs/electronicversions/pdfs/lglb106.pdf (last visited Feb. 20, 2008). The author discusses constitutional challenges to breed-specific legislation in other jurisdictions, with a specific focus on challenges based on vagueness, equal protection, and due process. She recommends that North Carolina jurisdictions interested in adopting a breed-specific ordinance

- identify a rational basis for regulating the breed;
- list specific breeds regulated;
- provide a uniform standard for determining when the ordinance applies to a mixed-breed dog;
- create a procedure for dog owners to ask whether their dog falls within the ordinance;
- provide for civil, rather than criminal, sanctions; and
- provide a hearing to a dog's owner if and when the government intends to destroy his or her dog.

The author also discusses alternatives to breed-specific legislation, such as rigorous enforcement of state and local dangerous dog laws.

28. *Id.*

Relevant Statutes

Article 1A of Chapter 67
Dangerous Dogs.

§ 67-4.1. Definitions and procedures.

(a) As used in this Article, unless the context clearly requires otherwise and except as modified in subsection (b) of this section, the term:

(1) "Dangerous dog" means

a. A dog that:

1. Without provocation has killed or inflicted severe injury on a person; or
2. Is determined by the person or Board designated by the county or municipal authority responsible for animal control to be potentially dangerous because the dog has engaged in one or more of the behaviors listed in subdivision (2) of this subsection.

b. Any dog owned or harbored primarily or in part for the purpose of dog fighting, or any dog trained for dog fighting.

(2) "Potentially dangerous dog" means a dog that the person or Board designated by the county or municipal authority responsible for animal control determines to have:

a. Inflicted a bite on a person that resulted in broken bones or disfiguring lacerations or required cosmetic surgery or hospitalization; or

b. Killed or inflicted severe injury upon a domestic animal when not on the owner's real property; or

c. Approached a person when not on the owner's property in a vicious or terrorizing manner in an apparent attitude of attack.

(3) "Owner" means any person or legal entity that has a possessory property right in a dog.

(4) "Owner's real property" means any real property owned or leased by the owner of the dog, but does not include any public right-of-way or a common area of a condominium, apartment complex, or townhouse development.

(5) "Severe injury" means any physical injury that results in broken bones or disfiguring lacerations or required cosmetic surgery or hospitalization.

(b) The provisions of this Article do not apply to:
  (1) A dog being used by a law enforcement officer to carry out the law enforcement officer's official duties;
  (2) A dog being used in a lawful hunt;
  (3) A dog where the injury or damage inflicted by the dog was sustained by a domestic animal while the dog was working as a hunting dog, herding dog, or predator control dog on the property of, or under the control of, its owner or keeper, and the damage or injury was to a species or type of domestic animal appropriate to the work of the dog; or
  (4) A dog where the injury inflicted by the dog was sustained by a person who, at the time of the injury, was committing a willful trespass or other tort, was tormenting, abusing, or assaulting the dog, had tormented, abused, or assaulted the dog, or was committing or attempting to commit a crime.

(c) The county or municipal authority responsible for animal control shall designate a person or a Board to be responsible for determining when a dog is a "potentially dangerous dog" and shall designate a separate Board to hear any appeal. The person or Board making the determination that a dog is a "potentially dangerous dog" must notify the owner in writing, giving the reasons for the determination, before the dog may be considered potentially dangerous under this Article. The owner may appeal the determination by filing written objections with the appellate Board within three days. The appellate Board shall schedule a hearing within 10 days of the filing of the objections. Any appeal from the final decision of such appellate Board shall be taken to the superior court by filing notice of appeal and a petition for review within 10 days of the final decision of the appellate Board. Appeals from rulings of the appellate Board shall be heard in the superior court division. The appeal shall be heard de novo before a superior court judge sitting in the county in which the appellate Board whose ruling is being appealed is located.

§ 67-4.2. Precautions against attacks by dangerous dogs.

(a) It is unlawful for an owner to:
  (1) Leave a dangerous dog unattended on the owner's real property unless the dog is confined indoors, in a securely enclosed and locked pen, or in another structure designed to restrain the dog;
  (2) Permit a dangerous dog to go beyond the owner's real property unless the dog is leashed and muzzled or is otherwise securely restrained and muzzled.

(b) If the owner of a dangerous dog transfers ownership or possession of the dog to another person (as defined in G.S. 12-3(6)), the owner shall provide written notice to:

(1) The authority that made the determination under this Article, stating the name and address of the new owner or possessor of the dog; and

(2) The person taking ownership or possession of the dog, specifying the dog's dangerous behavior and the authority's determination.

(c) Violation of this section is a Class 3 misdemeanor.

§ 67-4.3. Penalty for attacks by dangerous dogs.
The owner of a dangerous dog that attacks a person and causes physical injuries requiring medical treatment in excess of one hundred dollars ($100.00) shall be guilty of a Class 1 misdemeanor.

§ 67-4.4. Strict liability.
The owner of a dangerous dog shall be strictly liable in civil damages for any injuries or property damage the dog inflicts upon a person, his property, or another animal.

§ 67-4.5. Local ordinances.
Nothing in this Article shall be construed to prevent a city or county from adopting or enforcing its own program for control of dangerous dogs.

---

§ 130A-200. Confinement or leashing of vicious animals.
A local health director may declare an animal to be vicious and a menace to the public health when the animal has attacked a person causing bodily harm without being teased, molested, provoked, beaten, tortured or otherwise harmed. When an animal has been declared to be vicious and a menace to the public health, the local health director shall order the animal to be confined to its owner's property. However, the animal may be permitted to leave its owner's property when accompanied by a responsible adult and restrained on a leash.

Chapter 5

Nuisance and At-Large Animals

Local animal control officials spend much of their time dealing with citizen complaints about nuisance animals and animals that are found wandering at large. Although a handful of narrowly tailored state laws address at-large animals, local animal officers' primary enforcement activities involve local nuisance ordinances. These ordinances vary tremendously in scope and in their level of detail; a few focus primarily on at-large animals and sanitation, while others define *nuisance* broadly to encompass a wide range of animal issues.

State Law

North Carolina has no statewide leash law or nuisance statute. The rabies law does provide a partial mechanism for addressing the issue of stray animals; as discussed in Chapter 4, animal control officers may impound unidentified cats and dogs that are not wearing rabies tags.[1] If, however, the cat or dog at large is wearing its rabies tag, animal control officials have limited options under state law. One statute makes it a Class 3 misdemeanor to intentionally, knowingly, and willfully allow a dog over six months of age to run at large in the nighttime.[2] Several other narrowly worded state laws relate to specific nuisances and at-large animals.

- A person who owns or has possession of a female dog and *knowingly* permits the dog to run at large during estrus (i.e., "in heat" or "in season") may be charged with a Class 3 misdemeanor.[3]
- A person may be charged with a Class 3 misdemeanor if he or she (1) owns a dog that kills a domestic animal or person; (2) refuses to put the dog to death; and (3) permits the dog to "go at liberty." In addition to defining the

---

1. N.C. Gen. Stat. 130A-192 (hereinafter G.S.).
2. G.S. 67-12.
3. G.S. 67-2.

criminal charge, the law authorizes anyone finding the dog at large to kill it.[4]

- An owner or keeper whose dog, while off its home premises, kills or injures any livestock or fowl is liable for damages to the owner of the livestock or fowl.[5]
- Finally, state law includes specific provisions governing unauthorized dogs in wildlife areas. If a dog is found unmuzzled and running at large in a wildlife refuge, sanctuary, or management area, wildlife officials may seize and impound the dog.[6] They are also authorized to destroy humanely any dog found tracking, running, injuring, or killing a deer or bear during a season in which hunting with dogs is prohibited.[7]

Local Ordinances

Many jurisdictions have adopted ordinances governing nuisance animals and animals that are found running at large. Cities and counties may rely on three sources of statutory authority for these laws. First, both have broad police powers that give them the authority to adopt ordinances to "define, prohibit, regulate, or abate acts, omissions, or conditions detrimental to the health, safety or welfare of [their] citizens and the peace and dignity" of the city or county and to "define and abate nuisances."[8] Second, they have the authority to "regulate, restrict, or prohibit the possession or harboring . . . of animals which are dangerous to persons or property."[9] Finally, cities have specific authority to "regulate, restrict, or prohibit the keeping, running, or going at-large of any domestic animals, including dogs and cats."[10] By relying on this combination of statutes, local governments have been able to regulate many animal-related nuisances within their jurisdictions.

Some types of nuisances may be creatively addressed through other types of ordinances. For example, some jurisdictions rely on zoning ordinances to restrict the number of animals allowed in residential areas.[11]

---

4. GS. 67-3.
5. G.S. 67-1.
6. G.S. 67-14.1(b).
7. G.S. 67-14.1(a).
8. G.S. 153A-121(a) (counties); G.S. 160A-174(a) (cities).
9. G.S. 153A-131 (counties); G.S. 160A-187 (cities).
10. G.S. 160A-186.
11. *See, e.g.,* Cumberland County Zoning Ordinance, § 912 (restricting kennel operations in residential districts); *id.*, § 203 (defining *kennel* as a premises where four or more adult dogs are kept commercially or as pets, subject to limited exceptions).

The term *public nuisance* is commonly understood to mean a behavior or condition that unreasonably interferes with rights common to the general public.[12] Rather than rely on this general legal concept, most animal control ordinances include a definition or application of the nuisance that is more specifically focused on animals. Many ordinances address such commonly encountered animal behaviors as

- running at large,
- barking, howling, or making other noises,
- chasing or snapping at people or vehicles,
- turning over garbage receptacles, and
- damaging real or personal property.

Some nuisance ordinances also concern the care and keeping of animals, such as

- keeping animals too close to a right-of-way,
- housing animals in unsanitary conditions,
- failing to confine a female animal in estrus,
- failing to remove animal feces from public or private property, and
- keeping certain types of animals, such as poultry and livestock, within city limits.

With some exceptions, local animal nuisance ordinances across the state promote similar policy goals and use comparable language. Some variation occurs, though, with respect to noise and sanitation. A number of animal noise ordinances include general prohibitions on "excessive," "continuous," or "untimely" barking, while others are more specific. For example,

- Sanford's ordinance prohibits barking during certain hours of the night (11:00 P.M. to 5:00 A.M.).[13]
- Durham County's ordinance defines nuisance to include any animal that "continuously barks, howls, whines or mews in an excessive manner (one or more times per minute, each minute, during a ten-minute period)."[14]
- Winston-Salem's ordinance governs dogs "which habitually and regularly bark, howl or whine for at least 15 minutes so as to result in serious annoyance to neighboring residents and [so] as to interfere with the reasonable use and enjoyment of the premises occupied by such residents."[15]

---

12. W. Keeton et al., Prosser and Keeton on the Law of Torts §§ 86, 87, at 618–19 (5th ed. 1984).
13. Sanford Code of Ordinances, § 4-36(b)(3).
14. Durham County Code of Ordinances, § 4-13.
15. Winston-Salem Code of Ordinances, § 6-9.

With respect to sanitation, or "pooper scooper," ordinances, jurisdictions typically take one of two approaches. They either establish a general prohibition on "soiling" or "defiling" private or public property[16] or specifically require the animal's custodian to remove feces deposited on public or private property unless the property owner has given permission for it to remain.[17] One jurisdiction, Apex, struggled with the enforcement of such an ordinance and adopted a slightly different approach. It requires the removal of feces but also requires that

> [a]ny person, while harboring, walking, in possession of or in charge of a dog on public property, public park property, public right-of-way or any private property without the permission of the private property owner, shall have in his or her possession a bag or other container that closes, which is suitable for removing feces deposited by the dog.[18]

By requiring the custodian of the animal to have a bag or other container, the ordinance eliminates the need to either "catch the dog in the act" or find witnesses who did.

Challenges to local government authority to regulate nuisances have seldom been successful. The North Carolina Supreme Court has repeatedly found that local governments may regulate the keeping of animals pursuant to their general ordinance-making authority. In *State v. Harrell*, for example, the court recognized that "'ordinances regulating dogs and requiring them to be registered and licensed, and at times muzzled and prevented from going at-large, are within the police powers usually conferred upon the local corporation.'"[19]

Several cases have addressed specifically the issue of livestock running at large. Back in 1925, the court decided *State v. Stowe*, which held that Charlotte had the authority to regulate the keeping of cattle within the city limits.[20] In

---

16. *See, e.g.,* Forsyth County Code of Ordinances, § 6-1; Brevard Code of Ordinances, § 14-1; Surf City Code of Ordinances, § 3-1(2)(a).

17. *See, e.g.,* Garner Code of Ordinances, § 3-15(b)(7); Rocky Mount Code of Ordinances, § 4-22(a)(4) and (5).

18. Apex Code of Ordinances, § 4-1(b).

19. *Harrell*, 203 N.C. 210, 215, 165 S.E. 551, 553 (1932) (*quoting* EUGENE McQUILLIN, THE LAW OF MUNICIPAL CORPORATIONS , vol 3, § 1004 (2d ed.)).

20. *Stowe*, 190 N.C. 79, 128 S.E. 481 (1925). Earlier cases recognized the validity of such ordinances as well. *See, e.g.,* State v. Tweedy, 115 N.C. 704, 20 S.E. 183 (1894) (holding that a person could not be convicted of killing livestock running at large when the livestock was running at large in violation of a valid local ordinance); Jones v. Duncan, 127 N.C. 118, 37 S.E. 135 (1900) (confirming that the city had the authority to adopt an ordinance prohibiting livestock from running at large).

1983 the court reinforced its decision in *Stowe* when it upheld another livestock-related nuisance ordinance and explained that "prohibiting the keeping of animals other than house pets is reasonably related to the health and welfare of the citizens of the Town of Atlantic Beach."[21] Most recently, the court of appeals addressed the issue in the case of *State v. Lutterloah*. The Town of Carolina Beach adopted an ordinance that made it "unlawful to ride, lead, or drive any animal upon the sidewalk, boardwalk, roads, or beaches within the corporate limits."[22] The case arose after a police officer cited Lutterloah for violating the ordinance by riding a horse on a city street. In an unpublished decision, the court of appeals concluded that the ordinance was a valid exercise of the city's authority to regulate the "keeping, running, or going at-large" of animals.[23]

Ordinances targeting animal noise have also been upheld by the courts, despite the relatively general standards and language used in the laws. For example, Martin County's animal noise ordinance provides that "[i]t shall be unlawful for any person to own, keep, or have within the county an animal that habitually or repeatedly makes excessive noises that tend to annoy, disturb, or frighten its citizens."[24] The court rejected claims that the ordinance is unconstitutionally vague, or indefinite. It explained that while the terms used in the ordinance are general in nature, they have "commonly accepted meanings and are sufficiently certain to inform persons of ordinary intelligence as to what constitutes a violation."[25] The court further noted that such ordinances must be enforced and reviewed "based upon an objective standard" that considers the common meanings of the terms used in the law.[26]

While state law governing animal-related nuisances is sparse, local governments have expansive authority to address such concerns via ordinance. Courts are likely to uphold such ordinances as long as the government is able to point to specific statutory authority, such as the authority granted to cities to regulate domestic animals, or to articulate a rational reason why the law is an appropriate exercise of the government's police power.

---

21. Town of Atlantic Beach v. Young, 307 N.C. 422, 428, 298 S.E.2d 686, 691 (1983).
22. *Lutterloah*, 171 N.C.App. 516, 615 S.E.2d 738 (2005) (unpublished disposition).
23. *Id.* (citing G.S. 160A-186).
24. State v. Taylor, 128 N.C. App. 616, 619, 495 S.E.2d 413, 415 (1998).
25. *Id.*
26. *Id.* at 620, 495 S.E. at 416. *See also* State v. Dorsett, 3 N.C.App 331, 335–36, 164 S.E.2d 607, 610 (1968) (upholding a Greensboro noise ordinance that prohibited "unreasonably loud, disturbing and unnecessary noise in the city"); State v. Garren, 117 N.C.App. 393, 397–98, 451 S.E.2d 315, 318–19 (1994) (upholding a provision of a Jackson County noise ordinance prohibiting "loud, raucous and disturbing noise," which was defined as any sound that "annoys, disturbs, injures or endangers the comfort, health, peace or safety of reasonable persons of ordinary sensibilities").

Relevant Statutes

Article 1 of Chapter 67
Owner's Liability.

§ 67-1. Liability for injury to livestock or fowls.
If any dog, not being at the time on the premises of the owner or person having charge thereof, shall kill or injure any livestock or fowls, the owner or person having such dog in charge shall be liable for damages sustained by the injury, killing, or maiming of any livestock, and costs of suit.

§ 67-2. Permitting bitch at large.
If any person owning or having any bitch shall knowingly permit her to run at large during the erotic stage of copulation he shall be guilty of a Class 3 misdemeanor.

§ 67-3. Sheep-killing dogs to be killed.
If any person owning or having any dog that kills sheep or other domestic animals, or that kills a human being, upon satisfactory evidence of the same being made before any judge of the district court in the county, and the owner duly notified thereof, shall refuse to kill it, and shall permit such dog to go at liberty, he shall be guilty of a Class 3 misdemeanor, and the dog may be killed by anyone if found going at large.

. . .

Article 2 of Chapter 67
License Taxes on Dogs.

§ 67-12. Permitting dogs to run at large at night; penalty; liability for damage.
No person shall allow his dog over six months old to run at large in the nighttime unaccompanied by the owner or by some member of the owner's family, or some other person by the owner's permission. Any person intentionally, knowingly, and willfully violating this section shall be guilty of a Class 3 misdemeanor, and shall also be liable in damages to any person injured or suffering loss to his property or chattels.

§ 67-14.1. Dogs injuring deer or bear on wildlife management area may be killed; impounding unmuzzled dogs running at large.

(a) Any dog which trails, runs, injures or kills any deer or bear on any wildlife refuge, sanctuary or management area, now or hereafter so designated and managed by the Wildlife Resources Commission, during the closed season for hunting with dogs on such refuge or management area, is hereby declared to be a public nuisance, and any wildlife protector or other duly authorized agent or employee of the Wildlife Resources Commission may destroy, by humane method, any dog discovered trailing, running, injuring or killing any deer or bear in any such area during the closed season therein for hunting such game with dogs, without incurring liability by reason of his act in conformity with this section.

(b) Any unmuzzled dog running at large upon any wildlife refuge, sanctuary, or management area, when unaccompanied by any person having such dog in charge, shall be seized and impounded by any wildlife protector, or other duly authorized agent or employee of the Wildlife Resources Commission.

(c) The person impounding such dog shall cause a notice to be published at least once a week for two successive weeks in some newspaper published in the county wherein the dog was taken, or if none is published therein, in some newspaper having general circulation in the county. Such notice shall set forth a description of the dog, the place where it is impounded, and that the dog will be destroyed if not claimed and payment made for the advertisement, a catch fee of one dollar ($1.00) and the boarding, computed at the rate of fifty cents (50¢) per day, while impounded, by a certain date which date shall be not less than 15 days after the publication of the first notice. A similar notice shall be posted at the courthouse door.

(d) The owner of the dog, or his agent, may recover such dog upon payment of the cost of the publication of the notices hereinbefore described together with a catch fee of one dollar ($1.00) and the expense, computed at the rate of fifty cents (50¢) per day, incurred while impounding and boarding the dog.

(e) If any impounded dog is not recovered by the owner within 15 days after the publication of the first notice of the impounding, the dog may be destroyed in a humane manner by any wildlife protector or other duly authorized agent or employee of the North Carolina Wildlife Resources Commission, and no liability shall attach to any person acting in accordance with this section.

. . .

---

§ 130A-192. Dogs and cats not wearing required rabies vaccination tags. The Animal Control Officer shall canvass the county to determine if there are any dogs or cats not wearing the required rabies vaccination tag. If a dog or cat is found not wearing the required tag, the Animal Control Officer shall check to see if the owner's identification can be found on the animal. If the animal is wearing an owner identification tag, or if the Animal Control Officer otherwise knows who the owner is, the Animal Control Officer shall notify the owner in writing to have the animal vaccinated against rabies and to produce the required rabies vaccination certificate to the Animal Control Officer within three days of the notification. If the animal is not wearing an owner identification tag and the Animal Control Officer does not otherwise know who the owner is, the Animal Control Officer may impound the animal. The duration of the impoundment of these animals shall be established by the county board of commissioners, but the duration shall not be less than 72 hours. During the impoundment period, the Animal Control Officer shall make a reasonable effort to locate the owner of the animal. If the animal is not reclaimed by its owner during the impoundment period, the animal shall be disposed of in one of the following manners: returned to the owner; adopted as a pet by a new owner; sold to institutions within this State registered by the United States Department of Agriculture pursuant to the Federal Animal Welfare Act, as amended; or put to death by a procedure approved by the American Veterinary Medical Association, the Humane Society of the United States or of the American Humane Association. The Animal Control Officer shall maintain a record of all animals impounded under this section which shall include the date of impoundment, the length of impoundment, the method of disposal of the animal and the name of the person or institution to whom any animal has been released.

§ 153A-121. General ordinance-making power.

(a) A county may by ordinance define, regulate, prohibit, or abate acts, omissions, or conditions detrimental to the health, safety, or welfare of its citizens and the peace and dignity of the county; and may define and abate nuisances.

(b) This section does not authorize a county to regulate or control vehicular or pedestrian traffic on a street or highway under the control of the Board of Transportation, nor to regulate or control any right-of-way or right-of-passage belonging to a public utility, electric or telephone membership corporation, or public agency of the State. In addition, no county ordinance may regulate or control a highway right-of-way in a manner inconsistent with State law or an ordinance of the Board of Transportation.

(c) This section does not impair the authority of local boards of health to adopt rules and regulations to protect and promote public health.

§ 153A-131. Possession or harboring of dangerous animals.

A county may by ordinance regulate, restrict, or prohibit the possession or harboring of animals which are dangerous to persons or property. No such ordinance shall have the effect of permitting any activity or condition with respect to a wild animal which is prohibited or more severely restricted by regulations of the Wildlife Resources Commission.

§ 160A-174. General ordinance-making power.

(a) A city may by ordinance define, prohibit, regulate, or abate acts, omissions, or conditions, detrimental to the health, safety, or welfare of its citizens and the peace and dignity of the city, and may define and abate nuisances.

(b) A city ordinance shall be consistent with the Constitution and laws of North Carolina and of the United States. An ordinance is not consistent with State or federal law when:

(1) The ordinance infringes a liberty guaranteed to the people by the State or federal Constitution;
(2) The ordinance makes unlawful an act, omission or condition which is expressly made lawful by State or federal law;
(3) The ordinance makes lawful an act, omission, or condition which is expressly made unlawful by State or federal law;
(4) The ordinance purports to regulate a subject that cities are expressly forbidden to regulate by State or federal law;
(5) The ordinance purports to regulate a field for which a State or federal statute clearly shows a legislative intent to provide a complete and integrated regulatory scheme to the exclusion of local regulation;
(6) The elements of an offense defined by a city ordinance are identical to the elements of an offense defined by State or federal law.

The fact that a State or federal law, standing alone, makes a given act, omission, or condition unlawful shall not preclude city ordinances requiring a higher standard of conduct or condition.

§ 160A-186. Regulation of domestic animals.

A city may by ordinance regulate, restrict, or prohibit the keeping, running, or going at large of any domestic animals, including dogs and cats. The ordinance may provide that animals allowed to run at large in violation of the ordinance may be seized and sold or destroyed after reasonable efforts to notify their owner.

§ 160A-187. Possession or harboring of dangerous animals.

A city may by ordinance regulate, restrict, or prohibit the possession or harboring within the city of animals which are dangerous to persons or property. No such ordinance shall have the effect of permitting any activity or condition with respect to a wild animal which is prohibited or more severely restricted by regulations of the Wildlife Resources Commission.

Chapter 6

Exotic Animals

In North Carolina, private ownership of inherently dangerous, non-native, or exotic animals ("exotic animals") is regulated primarily by local governments. This chapter briefly reviews the arguments for and against such regulation, summarizes the limited provisions that address exotic animals in federal and state law, and analyzes some of the ordinances adopted by the state's cities and counties.

Background

Private ownership of exotic animals, like many animal-related issues, is quite controversial. Proponents of exotic animal regulation typically offer three justifications.[1] First, they argue, introduction of exotic species can present a risk to public safety. In 2003, for example, a ten-year old Wilkes County boy was killed by his aunt's pet tiger, and in 2004 a fourteen-year old girl in Surry County was attacked by one of her family's four pet tigers.[2] Second, say proponents of regulation, exotic animals can create threats to both the public health and the environment.[3] Recent examples of such risks include the 2003 outbreak of

---

1. *See* Matthew G. Liebman, "Detailed Discussion of Exotic Pet Laws" (2004), Animal Legal and Historical Center, www.animallaw.info/articles/ddusexoticpets.htm (last visited Feb. 25, 2008).

2. Dennis Rogers, "Time for a Ban on Big Cats," *Raleigh News & Observer*, Feb. 14, 2004, www.newsobserver.com/news/rogers/2004/story/168793.html (last visited Feb. 25, 2008).

3. *See, e.g.,* Robert A. Cook, "Importation of Exotic Species and the Impact on Public Health and Safety," Testimony on behalf of the Wildlife Conservation Society and the American Zoo and Aquarium Association before the U. S. Senate Committee on Environment and Public Works (July 17, 2003), http://www.aza.org/RC/Documents/TestimonyZoonoses.pdf (last visited Sept. 24, 2007); Gabriela Chavarria, "Importation of Exotic Species," Testimony on behalf of the National Environmental Coalition on

monkeypox in humans attributed to pet prairie dogs[4] and the escape and proliferation of Burmese pythons in the Florida Everglades.[5] Finally, proponents argue, exotic animals often suffer when owned privately because their keepers are neither trained nor equipped to provide appropriate care.[6]

Opponents argue that regulation interferes with the educational benefits associated with private zoos and traveling animal exhibitions.[7] Others contend that private owners can help prevent the extinction of many species through captive breeding programs.[8] Finally, some disagree with the public health and safety arguments put forward by proponents and argue that government regulation unnecessarily infringes individuals' property rights to own such animals and earn a living through private enterprises.[9]

Federal Law

Several federal agencies are involved in regulating exotic animals. Public health officials have issued several animal-related regulations over the years in response to communicable disease threats, and wildlife officials have exercised their relatively expansive authority to regulate the introduction of exotic species into the country.

Invasive Species before the U.S. States Senate Committee on Environment and Public Works (July 17, 2003), http://epw.senate.gov/hearing_statements.cfm?id=212884 (last visited Feb. 25, 2008).

4. Centers for Disease Control and Prevention, "Update: Multistate Outbreak of Monkeypox—Illinois, Indiana, Kansas, Missouri, Ohio, and Wisconsin, 2003," *Mortality and Morbidity Weekly Report* 52 (July 11, 2003): 642–46, www.cdc.gov/mmwr/PDF/wk/mm5227.pdf (last visited Feb. 25, 2008).

5. *See* Robert Brown, "Note: Exotic Pets Invade United States Ecosystems: Legislative Failure and a Proposed Solution," *Indiana Law Journal* 81 (2006): 713; *see also* Everglades Burmese Python Project (a research and assistance project offered through Davidson College's herpetology laboratory), www.bio.davidson.edu/people/midorcas/research/StResearch/Python%20Project%20Website/Python.htm (last visited Feb. 25, 2008).

6. *See* Liebman, "Exotic Pet Laws"; Animal Protection Institute, "A Life Sentence: The Sad and Dangerous Realities of Exotic Animals in Private Hands," www.api4animals.org/a3b_exotic_pets.php (last visited Feb. 25, 2008).

7. Jim Nesbitt, "Who Should Keep Exotic Animals?" *Raleigh News & Observer*, July 8, 2007, 24A.

8. Zuzana Kukol, "Questions and Answers about Keeping Wild and Exotic Animals in Captivity" (August 2007), www.rexano.org/Documents/Wild_Exotic_Pet_Animal.pdf (last visited Feb. 25, 2008).

9. *Id.*

The Food and Drug Administration (FDA) in the U.S. Department of Health and Human Services has regulations in place that (1) prohibit the sale of turtles (with shells four inches long or smaller) and turtle eggs[10] and (2) restrict interstate transport of psittacine birds, such as parrots, cockatoos, and parakeets.[11] These regulations are intended to minimize the transmission of salmonellosis (from turtles) and psittacosis (from birds) to humans.

In addition, following the 2003 monkeypox outbreak, the FDA and the Centers for Disease Control and Prevention adopted regulations prohibiting the import of African rodents (or products from such rodents) and the capture, sale, transport, or release of several animals with known potential for transmitting the monkeypox virus to humans; the prohibited animals include prairie dogs, Gambian rats, and certain squirrels and porcupines.[12] Both sets of monkeypox regulations provide for limited exceptions to the prohibitions.

The U.S. Fish and Wildlife Service of the Department of the Interior exercises regulatory authority that extends beyond the communicable disease realm. It is responsible for implementing federal laws that make it unlawful to

- import animals and fish the agency deems "injurious" to people or agricultural interests;[13]
- import, export, transport, sell, receive, acquire, or purchase fish, wildlife, or plants that are taken, possessed, transported, or sold in violation of any law (including foreign, federal, and state laws);[14] and
- import, export, transport, sell, receive, acquire, or purchase live lions, tigers, leopards, snow leopards, clouded leopards, cheetahs, jaguars, or cougars, or any hybrid combination of these species.[15]

A person who violates one of these laws may be subject to civil and criminal penalties.[16]

There are some limited exceptions to these federal laws. For example, wildlife rehabilitators, universities, veterinarians, and certain wildlife sanctuaries are not subject to the same restrictions as the general public. Under some

---

10. 21 C.F.R. 1240.62.
11. 21 C.F.R. 1240.65.
12. 21 C.F.R. 1240.63 (FDA regulation); 42 C.F.R. 71.56.
13. 18 U.S.C. § 42.
14. The Lacey Act, 16 U.S.C. §§ 3371–3378.
15. 16 U.S.C. §§ 3371–3372; 50 U.S.C. 14.250 to 14.255. The provisions of the law addressing large cats are known as the Captive Wildlife Safety Act, Pub. L. 108-191, amending the Lacey Act, 16 U.S.C. §§ 3371–3372.
16. 16 U.S.C. § 3374.

circumstances, the Fish and Wildlife Service may also issue permits allowing the importation of animals considered injurious.[17]

State Law

North Carolina does not have a general statewide law regulating the ownership or possession of exotic or dangerous animals. According to the Animal Protection Institute, an advocacy organization that supports regulation of private ownership of exotic animals,

- eighteen states have a ban on private ownership;
- ten states have a partial ban on private ownership;
- thirteen states require owners of exotic animals to obtain a license or permit from the state; and
- nine states, including North Carolina, have not adopted a state law banning private ownership or requiring licenses or permits.[18]

In 2006 the General Assembly directed a study commission to identify animals that are so dangerous that they should not be owned or possessed by private individuals and to offer recommendations regarding appropriate state regulation of the keeping of such animals. [19] The study commission offered its conclusions and recommendations to the General Assembly in May 2007. It identified various species that should be regulated by the state as inherently dangerous animals, recommended a system of grandfathering for current owners of such animals, and suggested the types of people and organizations that should be exempt from state regulation.[20] The 2007 General Assembly considered but did not act on legislation that incorporated some of the study commission's recommendations.[21]

While North Carolina does not have a comprehensive state law, it does have laws that regulate native wildlife as well as the ownership, possession, or keep-

---

17. Permit information is available at www.fws.gov/permits/instructions/ObtainPermit.shtml.

18. Animal Protection Institute (API), State Laws Governing Private Possession of Exotic Animals. API recognizes that some of these last nine states may regulate entry of exotic animals into the state or require a veterinary certificate. North Carolina does not have any such regulation, www.api4animals.org/b4a2_exotic animals_map.php (last visited Feb. 25, 2008).

19. S.L. 2006-248, sec. 32.1 through 32.3.

20. Inherently Dangerous Exotic Animals in North Carolina: Recommendations of the Study Committee (May 2007) (on file with author).

21. S 1477/H 1614.

ing of certain exotic or dangerous animals. The state, through the Wildlife Resources Commission (WRC), exercises jurisdiction over native North Carolina wildlife and regulates the killing or capture of such animals.[22] In addition, the WRC has the authority to regulate the "acquisition, importation, possession, transportation, disposition, or release into public or private waters or the environment of zoological or botanical species or specimens that may threaten the introduction of epizootic disease or may create a danger to or an imbalance in the environment inimical to the conservation of wildlife resources."[23] This WRC regulatory authority is limited, however, to species or specimens that may have a negative environmental impact. It does not appear to extend to regulation of animals that some perceive as dangerous for other reasons, such as tigers.

With respect to non-native wildlife, the state has statutes that make it unlawful to

- intentionally expose humans to venomous reptiles.[24]
- raise American alligators without a proper license.[25]
- release exotic species of wild animals or wild birds into an area for the purpose of stocking an area for hunting or trapping.[26]
- sell or barter turtles.[27]

In addition, local public health directors have independent authority to declare "vicious and a menace to the public health" any animal that makes an unprovoked attacked on a person causing bodily harm.[28] Once an animal has been declared vicious, it must be confined to its owner's property except when (1) accompanied by a responsible adult and (2) restrained on a leash.

---

22. *See* Article 24 of N.C. GEN. STAT. Chapter 143 (hereinafter G.S.) (establishing the Wildlife Resources Commission); G.S. Ch. 113 (including various statutes governing the jurisdiction and activities of the commission).

23. G.S. 113-292(d).

24. G.S. 14-416 et seq.

25. G.S. 106-763.1.

26. G.S. 113-292(e).

27. N.C. ADMIN. CODE tit. 10A, ch. 41A, § .0302 (hereinafter N.C.A.C.). This regulation is intended to "prevent the spread of salmonellosis from pet turtles to humans." The prohibition does not apply to sales of turtles used for scientific, educational, or food purposes. *Id.*

28. G.S. 130A-200. The animal must not have been "teased, molested, provoked, beaten, tortured, or otherwise harmed."

Local Ordinances

Many North Carolina local governments have adopted local exotic animal ordinances. According to one report, at least twenty-six counties and eleven municipalities have exotic animal laws in place.[29] Cities and counties interested in adopting such ordinances may rely upon two different sources of statutory authority.

- General authority to protect the health, safety, and welfare[30]
- Specific authority to regulate animals that are dangerous to persons or property[31]

The general ordinance-making authority, often referred to as "police power," is broad in scope and can be the basis for various types of exotic animal regulations, including a complete ban on private ownership. The second source of authority, which allows regulation of dangerous animals, may be interpreted narrowly or broadly. A narrow interpretation of the term *dangerous* would limit it to animals that might injure a person (e.g., tigers, venomous reptiles); but it could also be interpreted to extend to regulation of animals—such as prairie dogs—that are not necessarily threats to physical safety but may introduce diseases dangerous to humans or domestic animals.

The scope and content of local ordinances vary across the state. Based on a sample of ordinances, it appears that local governments considering regulation in this area must answer two closely related policy questions.[32]

- What is the ordinance trying to achieve?
- What types of animals does the jurisdiction want to regulate?

Local governments across the state have answered these questions differently.

What Is the Ordinance Trying to Achieve?

Local ordinances in North Carolina take a variety of different approaches to regulation of exotic animals within their jurisdictions. The most common tactic appears to be a ban on private ownership. For example:

---

29. Jim Nesbitt, "Who Should Keep Exotic Animals?" *Raleigh News & Observer*, July 8, 2007, 23A.

30. G.S. 153A-121 (counties); 160A-174 (cities).

31. G.S. 153A-131 (counties); 160A-187 (cities).

32. This section is based on a review of ordinances available on www.municode.com and on individual websites of some jurisdictions.

- "No person, firm, or corporation shall keep, maintain, possess or have within the county any venomous reptile or any other wild or exotic animal."[33]
- "It is unlawful to keep or harbor or breed or sell or trade any wild or exotic animal as a pet, for display or for exhibition purposes, whether gratuitously or for a fee, except as may be licensed by the state wildlife resources commission under its regulations pertaining to wildlife rehabilitators."[34]

Ordinances often include at least one exception to the ban. Examples of exceptions include circuses, carnivals, zoos. pet shops, animal transport vehicles passing through the jurisdiction, scientific research laboratories, veterinary clinics, and wildlife rehabilitators.

Some jurisdictions do not adopt bans on private ownership but take alternative approaches, including

- *Reporting requirements:* Owners of exotic animals in Cary or Smithfield must comply with several reporting requirements. They must notify animal control officials of (1) the animal's arrival; (2) any injuries to persons, other animals, or property caused by the animal; and (3) if the animal is required to be confined, any incident in which it escapes or roams at large.[35]
- *Permitting requirements:* Wilmington requires owners of livestock and wild animals to obtain permits, but it appears to limit wild animal permits to people who hold licenses or permits from the North Carolina Wildlife Resources Commission, such as wildlife captivity licenses or wildlife rehabilitation permits.[36]
- *Bonding requirements:* If a person wants to temporarily exhibit a wild or nondomesticated animal within the city of Henderson, he or she must post a $10,000 bond and have a $1 million insurance policy in place insuring against any damage or injury caused by the animal.[37]
- *Enclosures:* Orange County requires primary and secondary enclosures for wild animals.[38]

---

33. Buncombe County Code of Ordinances, § 6-61; *see also* Charlotte Code of Ordinances, § 3-73(a) (using similar language); Currituck County Code of Ordinances, § 3-88.
34. Fayetteville Code of Ordinances, § 6-73(b).
35. Cary Code of Ordinances, § 6-63; Smithfield Code of Ordinances, § 4-9(b).
36. Wilmington Code of Ordinances, § 6-2.1.
37. Henderson Code of Ordinances, § 6-10.
38. Orange County Code of Ordinances, § 4-133.

What Types of Animals Are Regulated?

Ordinances address and define a variety of different categories of animals, including those that are wild, exotic, and/or inherently dangerous. Some define a category by listing common names or species of the animals, while others define a category much more broadly. A few localities, such as Iredell County, not only define terms but include a list of animals that are *not* encompassed by the definition. A few examples of defined categories follow:

- *Exotic animals*: Exotic animals are animals other than domestic animals, farm animals, and wild animals which are not native to North Carolina.[39]
- *Wild animals* shall include an animal that (i) typically is found in a nondomesticated state and that, because of its size or vicious propensity or because it is poisonous or for any other substantial reason poses a potential danger to persons, other animals or property, or (ii) is classified as a wild animal by the North Carolina Wildlife Resources Commission (WRC) so that any person wishing to possess the same is required by state law to obtain a permit from WRC.[40]
- *Exotic or wild animal:* An animal that would ordinarily be confined to a zoo, or one that would ordinarily be found in the wilderness of this or any other country or one that is a species of animal not indigenous to the United States or to North America, or one that otherwise is likely to cause a reasonable person to be fearful of significant destruction of property or of bodily harm and the latter includes but is not limited to: monkeys, raccoons, squirrels, ocelots, bobcats, wolves, hybrid wolves, venomous reptiles, and other such animals. Such animals are further defined as those mammals or non-venomous reptiles weighing over fifty (50) pounds at maturity, which are known at law as ferae naturae. Exotic or wild animals specifically do not include animals of a species customarily used in North Carolina as ordinary household pets, animals of a species customarily used in North Carolina as domestic farm animals, fish confined in an aquarium other than piranha, birds, or insects.[41]
- (1) *Wild animals dangerous to humans and property.* Wild animals are any animals not normally domesticated. For purposes of this chapter, wild animals are deemed inherently dangerous. They are deemed as such because of their vicious propensities and capabilities, the likely gravity

39. Chapel Hill Code of Ordinances, § 4-1(k).

40. Wilmington Code of Ordinances, § 6-2.1.

41. Lexington Code of Ordinances, § 5-2. Similar definitions are used in other jurisdictions: *see, e.g.*, Charlotte Code of Ordinances, § 3-3; Davidson Code of Ordinances, § 10-31, Holly Springs Code of Ordinances, § 12-61.

of harm inflicted by their attack and unpredictability despite attempts at domestication. The category of wild animals includes but is not limited to:
a. Members of the Canidae family such as wolves, coyotes, and hybrids of those breeds.
b. Members of the Ursidae family which includes any member of the bear family or hybrids thereof.
c. Members of the Felidae family such as wild cats, cougars, mountain lions, or panthers.

- (2) *Exotic animals dangerous to humans and property.* Exotic animals are also considered to be inherently dangerous for purposes of this chapter. Like wild animals, exotic animals are dangerous because of their vicious propensities and capabilities, the gravity of harm inflicted by their attack, and unpredictability despite attempts at domestication. The category of exotic animals includes, but is not limited to:
a. Reptiles which are poisonous or constricting reptiles more than ten (10) feet in length.
b. Nonhuman primates weighing greater than twenty-five (25) pounds.
c. Members of the feline family other than domestic house cats, including, but not limited to lions, tigers and leopards.
d. Reptiles which are members of the crocodile family, including, but not limited to alligators and crocodiles.

Wild and exotic animals do not include:

(1) Wolf–dog cross breeds.
(2) Foreign rodents such as guinea pigs, hamsters, ferrets, and chinchillas.
(3) Members of the reptile and amphibian family not specifically mentioned above such as small lizards and iguanas, salamanders, turtles and frogs.
(4) Vietnamese potbellied pigs, and other members of the Suidae family, except wild boar and peccary.
(5) Horses and other members of the Equidae family.
(6) Cows and other members of the Bovidae family.
(7) Deer and other members of the Cervidae family.
(8) Domestic dogs and cats.

These animals do not have dangerous propensities and pose no serious threat to the safety of persons and property within Iredell County.[42]

One of the most controversial definitional issues is whether to include hybrid animals, particularly wolf–dog hybrids. Members of the state study commission that examined this issue in 2007 could not reach consensus on whether

42. Iredell County Code of Ordinances, § 3-1.

these animals were "inherently dangerous."[43] For the most part, North Carolina localities that regulate exotic animals have included wolf–dog hybrids within the scope of their ordinances, and several local governments have specifically included hybrids of other animals (such as bears and cats) as well.[44]

Conclusion

Taken together, state and federal laws and local ordinances create a patchwork of regulation governing some aspects of private ownership of wild and exotic animals. Given the national trend toward statewide regulation and the General Assembly's recent discussion about inherently dangerous animals, it would not be surprising if North Carolina implemented more comprehensive legislation in the coming years.

---

43. Inherently Dangerous Exotic Animals in North Carolina: Recommendations of the Study Committee (May 2007) (on file with author).

44. *See, e.g.*, Smithfield Code of Ordinances, § 4-9(a) ("A hybrid of any animal as defined in this section, regardless of genetic percentages, shall be deemed exotic."); Fayetteville Code of Ordinances, § 6-73 ("Wild or exotic animal means an animal which is usually not a domestic animal and which can normally be found in the wild state, including, but not limited to, lions, tigers, leopards, panthers, wolves, foxes, lynxes, or any hybrid of like animals, alligators, crocodiles, apes, foxes, elephants, rhinoceroses, bears, all forms of poisonous snakes, raccoons, skunks, monkeys, bats and like animals.").

Relevant Statutes

Article 55 of Chapter 14
Handling of Poisonous Reptiles.

§ 14-416. Handling of poisonous reptiles declared public nuisance and criminal offense.
The intentional exposure of human beings to contact with reptiles of a venomous nature being essentially dangerous and injurious and detrimental to public health, safety and welfare, the indulgence in and inducement to such exposure is hereby declared to be a public nuisance and a criminal offense, to be abated and punished as provided in this Article.

§ 14-417. Regulation of ownership or use of poisonous reptiles.
It shall be unlawful for any person to own, possess, use, or traffic in any reptile of a poisonous nature whose venom is not removed, unless such reptile is at all times kept securely in a box, cage, or other safe container in which there are no openings of sufficient size to permit the escape of such reptile, or through which such reptile can bite or inject its venom into any human being.

§ 14-418. Prohibited handling of reptiles or suggesting or inducing others to handle.
It shall be unlawful for any person to intentionally handle any reptile of a poisonous nature whose venom is not removed, by taking or holding such reptile in bare hands or by placing or holding such reptile against any exposed part of the human anatomy, or by placing their own or another's hand or any other part of the human anatomy in or near any box, cage, or other container wherein such reptile is known or suspected to be. It shall also be unlawful for any person to intentionally suggest, entice, invite, challenge, intimidate, exhort or otherwise induce or aid any person to handle or expose himself to any such poisonous reptile in any manner defined in this Article.

§ 14-419. Investigation of suspected violations; seizure and examination of reptiles; disposition of reptiles.
In any case in which any law-enforcement officer or animal control officer has reasonable grounds to believe that any of the provisions of this Article have been or are about to be violated, it shall be the duty of such officer and he is hereby authorized, empowered, and directed to immediately investigate such violation or impending violation and to forthwith seize the reptile or reptiles involved, and all such officers are hereby authorized and directed to deliver such reptiles to the North Carolina State Museum of Natural Sciences or to its

designated representative for examination and test for the purpose of ascertaining whether said reptiles contain venom and are poisonous. If the North Carolina State Museum of Natural Sciences or its designated representative finds that said reptiles are dangerously poisonous, the North Carolina State Museum of Natural Sciences or its designated representative shall be empowered to dispose of said reptiles in a manner consistent with the safety of the public; but if the Museum or its designated representative find that the reptiles are not dangerously poisonous, and are not and cannot be harmful to human life, safety, health or welfare, then it shall be the duty of such officers to return the said reptiles to the person from whom they were seized within five days.

§ 14-420. Arrest of persons violating provisions of Article.
If the examination and tests made by the North Carolina State Museum of Natural Sciences or its designated representative as provided herein show that such reptiles are dangerously poisonous, it shall be the duty of the officers making the seizure, in addition to destroying such reptiles, also to arrest all persons violating any of the provisions of this Article.

§ 14-421. Exemptions from provisions of Article.
This Article shall not apply to the possession, exhibition, or handling of reptiles by employees or agents of duly constituted museums, laboratories, educational or scientific institutions in the course of their educational or scientific work.

§ 14-422. Violation made misdemeanor.
Any person violating any of the provisions of this Article shall be guilty of a Class 2 misdemeanor.

---

§ 106-763.1. Propagation and production of American alligators.

(a) License Required. – A person who intends to raise American alligators commercially must first obtain an Aquaculture Propagation and Production Facility License from the Department. The Board of Agriculture may regulate a facility that raises American alligators to the same extent that it can regulate any other facility licensed under this Article.

(b) Requirements. – A facility that raises American alligators commercially must comply with all of the following requirements:

(1) Before a facility begins operation, it must prepare and implement a confinement plan. After a facility begins operation, it must adhere to the confinement plan. A confinement plan must comply with guidelines developed

and adopted by the Wildlife Resources Commission. The Department may inspect a facility to determine if the facility is complying with the confinement plan. As used in this subdivision, "confinement" includes production within a building or similar structure and a perimeter fence.

(2) A facility can possess only hatchlings that have been permanently tagged and have an export permit from their state of origin. The facility must keep records of all hatchlings it receives and must make these records available for inspection by the Wildlife Resources Commission and the Department upon request.

(3) If the facility uses swine, poultry, or other livestock for feed, it must have a disease management plan that has been approved by the State Veterinarian, and it must comply with the plan.

(4) The activities of the facility must comply with the Endangered Species Act and the Convention on International Trade in Endangered Species. The Department is the State agency responsible for the administration of this program for farm-raised alligators.

(c) Sanctions. – The operator of a facility that possesses an untagged or undocumented alligator commits a Class H felony if the operator knows the alligator is untagged or undocumented. Conviction of an operator of a facility under this section revokes the license of the facility for five years beginning on the date of the conviction. An operator convicted under this section may not be the operator of any other facility required to be licensed under this Article for five years beginning on the date of the conviction.

§ 113-292. Authority of the Wildlife Resources Commission in regulation of inland fishing and the introduction of exotic species.

(a) The Wildlife Resources Commission is authorized to authorize, license, regulate, prohibit, prescribe, or restrict all fishing in inland fishing waters, and the taking of inland game fish in coastal fishing waters, with respect to:

(1) Time, place, character, or dimensions of any methods or equipment that may be employed in taking fish;

(2) Seasons for taking fish;

(3) Size limits on and maximum quantities of fish that may be taken, possessed, bailed to another, transported, bought, sold, or given away.

(b) The Wildlife Resources Commission is authorized to authorize, license, regulate, prohibit, prescribe, or restrict:

(1) The opening and closing of inland fishing waters, whether entirely or only as to the taking of particular classes of fish, use of particular equipment, or as to other activities within the jurisdiction of the Wildlife Resources Commission; and
(2) The possession, cultivation, transportation, importation, exportation, sale, purchase, acquisition, and disposition of all inland fisheries resources and all related equipment, implements, vessels, and conveyances as necessary to implement the work of the Wildlife Resources Commission in carrying out its duties.

To the extent not in conflict with provisions enforced by the Department, the Wildlife Resources Commission may exercise the powers conferred in this subsection in coastal fishing waters pursuant to its rule of inland game fish in such waters.

(c) The Wildlife Resources Commission is authorized to make such rules pertaining to the acquisition, disposition, transportation, and possession of fish in connection with private ponds as may be necessary in carrying out the provisions of this Subchapter and the overall objectives of the conservation of wildlife resources.

(c1) The Wildlife Resources Commission is authorized to issue proclamations suspending or extending the hook-and-line season for striped bass in the inland and joint waters of coastal rivers and their tributaries, and the Commission may delegate this authority to the Executive Director. Each proclamation shall state the hour and date upon which it becomes effective, and shall be issued at least 48 hours prior to the effective date and time. A permanent file of the text of all proclamations shall be maintained in the office of the Executive Director. Certified copies of proclamations are entitled to judicial notice in any civil or criminal proceeding.

The Executive Director shall make reasonable effort to give notice of the terms of any proclamation to persons who may be affected by it. This effort shall include press releases to communications media, posting of notices at boating access areas and other places where persons affected may gather, personal communication by agents of the Wildlife Resources Commission, and other measures designed to reach persons who may be affected. Proclamations under this subsection shall remain in force until rescinded following the same procedure established for enactment.

(d) The Wildlife Resources Commission is authorized to authorize, license, regulate, prohibit, prescribe, or restrict anywhere in the State the acquisition, importation, possession, transportation, disposition, or release into public or private waters or the environment of zoological or botanical species or specimens that may threaten the introduction of epizootic disease or may create a danger to or an imbalance in the environment inimical to the conservation of wildlife resources. This subsection is not intended to give the Wildlife Resources Commission the authority to supplant, enact any conflicting rules, or otherwise take any action inconsistent with that of any other State agency acting within its jurisdiction.

(e) It is unlawful for any person to:

(1) Release or place exotic species of wild animals or wild birds in an area for the purpose of stocking the area for hunting or trapping;

(2) Release or place species of wild animals or wild birds not indigenous to that area in an area for the purpose of stocking the area for hunting or trapping;

(3) Take by hunting or trapping any animal or bird released or placed in an area in contravention of subdivisions (1) and (2) of this subsection, except under a permit to hunt or trap which may be issued by the Wildlife Resources Commission for the purpose of eradicating or controlling the population of any species of wildlife that has been so released or placed in the area.

§ 130A-200. Confinement or leashing of vicious animals.

A local health director may declare an animal to be vicious and a menace to the public health when the animal has attacked a person causing bodily harm without being teased, molested, provoked, beaten, tortured or otherwise harmed. When an animal has been declared to be vicious and a menace to the public health, the local health director shall order the animal to be confined to its owner's property. However, the animal may be permitted to leave its owner's property when accompanied by a responsible adult and restrained on a leash.

§ 153A-121. General ordinance-making power.

(a) A county may by ordinance define, regulate, prohibit, or abate acts, omissions, or conditions detrimental to the health, safety, or welfare of its citizens and the peace and dignity of the county; and may define and abate nuisances.

(b) This section does not authorize a county to regulate or control vehicular or pedestrian traffic on a street or highway under the control of the Board of Transportation, nor to regulate or control any right-of-way or right-of-passage belonging to a public utility, electric or telephone membership corporation, or public agency of the State. In addition, no county ordinance may regulate or control a highway right-of-way in a manner inconsistent with State law or an ordinance of the Board of Transportation.

(c) This section does not impair the authority of local boards of health to adopt rules and regulations to protect and promote public health.

§ 153A-131. Possession or harboring of dangerous animals.
A county may by ordinance regulate, restrict, or prohibit the possession or harboring of animals which are dangerous to persons or property. No such ordinance shall have the effect of permitting any activity or condition with respect to a wild animal which is prohibited or more severely restricted by regulations of the Wildlife Resources Commission.

§ 160A-174. General ordinance-making power.

(a) A city may by ordinance define, prohibit, regulate, or abate acts, omissions, or conditions, detrimental to the health, safety, or welfare of its citizens and the peace and dignity of the city, and may define and abate nuisances.

(b) A city ordinance shall be consistent with the Constitution and laws of North Carolina and of the United States. An ordinance is not consistent with State or federal law when:

(1) The ordinance infringes a liberty guaranteed to the people by the State or federal Constitution;
(2) The ordinance makes unlawful an act, omission or condition which is expressly made lawful by State or federal law;
(3) The ordinance makes lawful an act, omission, or condition which is expressly made unlawful by State or federal law;
(4) The ordinance purports to regulate a subject that cities are expressly forbidden to regulate by State or federal law;
(5) The ordinance purports to regulate a field for which a State or federal statute clearly shows a legislative intent to provide a complete and integrated regulatory scheme to the exclusion of local regulation;
(6) The elements of an offense defined by a city ordinance are identical to the elements of an offense defined by State or federal law.

The fact that a State or federal law, standing alone, makes a given act, omission, or condition unlawful shall not preclude city ordinances requiring a higher standard of conduct or condition.

§ 160A-187. Possession or harboring of dangerous animals.
A city may by ordinance regulate, restrict, or prohibit the possession or harboring within the city of animals which are dangerous to persons or property. No such ordinance shall have the effect of permitting any activity or condition with respect to a wild animal which is prohibited or more severely restricted by regulations of the Wildlife Resources Commission.

Chapter 7

Regulation of Animal Shelters, Kennels, and Other Operations

A wide range of people and organizations—such as shelters, research facilities, and pet stores—work with and assume responsibility for animals. Laws at both state and federal levels have evolved over time into a fairly complex body of regulation for these operations. The laws focus primarily on the welfare of the animals and rely on licensing and registration as their primary regulatory tools. This chapter reviews the state and federal laws in this area, with particular attention to the regulation of animal shelters operated by North Carolina local governments.

State Law

Local governments are allowed, but not required, to have animal shelters. They have the authority to either (1) "establish, equip, operate, and maintain" an animal shelter or (2) fund a shelter operated by another entity, such as an animal welfare organization.[1] Some local governments collaborate with neighboring jurisdictions to operate regional shelters, such as the Tri-County Animal Shelter that serves Chowan, Gates, and Perquimans counties.[2]

Until recently the state regulated only animal shelters run by private entities; those owned and operated by local governments were exempt.[3] In 2004 and

---

1. N.C. Gen. Stat. 153A-442 (counties) (hereinafter G.S.); G.S. 160A-493 (cities). Note that state law specifically authorizes local governments to enter into contracts with private entities "to carry out any public purpose that the [city/county] is authorized by law to engage in." G.S. 153A-449 (counties); 160A-20.1 (cities).

2. Limited information about the Tri-County Animal Shelter is available at www.petfinder.com/shelters/NC247.html (last visited Mar. 3, 2008).

3. Prior to the most recent change, the statutory definition of *animal shelter* was limited to facilities "owned, operated, or maintained by a duly incorporated humane society, animal welfare society, society for the prevention of cruelty to animals, or other nonprofit organization devoted to the welfare, protection and humane treatment of animals." G.S. 19A-23(5).

2005, however, the General Assembly passed legislation expanding the regulatory authority of the North Carolina Department of Agriculture and Consumer Services (the Department) to include local government shelters.[4] Under the state's Animal Welfare Act (NCAWA), the Department now regulates all the different types of facilities that house and, sometimes, sell animals.[5] The legislation explains that such regulation is necessary to

- protect pet owners from theft,
- prevent the sale or use of stolen pets,
- ensure that animals in commerce are provided humane care and treatment, and
- limit the sale, trade, or adoption of animals that show signs of being sick or having congenital abnormalities.[6]

In addition to animal shelters, the act regulates pet shops, public auctions, boarding kennels, and dealers. It exempts from regulation four categories of operations: (1) veterinary hospitals; (2) people who occasionally board animals on a noncommercial basis; (3) dealers, pet shops, public auctions, commercial kennels, or research facilities while they hold a license or registration pursuant to the federal Animal Welfare Act;[7] and (4) kennels establishments operated primarily for the boarding or training of hunting dogs.[8]

Registration and Licensure Requirements

Animal shelter operators must obtain certificates of registration from the Animal Welfare Section of the Department and renew those certificates each year.[9] The other four operations (pet shops, public auctions, boarding kennels, and

4. S.L. 2004-199 (amending the law to provide local governments the authority to establish, equip, operate, and maintain animal shelters [G.S. 153A-442; 16A-493] by adding the following sentence: "The animal shelters shall meet the same standards as animal shelters regulated" under the NCAWA); S.L. 2005-276, sec. 11.5 (Uniform Regulation of Animal Shelters) (amending the NCAWA to include public shelters within the definition of animal shelter). For further discussion of these legislative changes, see Aimee Wall, "North Carolina Animal Control Law: 2005 Legislative Update," *Local Government Law Bulletin* No. 107 (October 2005).

5. Article 3 of G.S. Ch. 19A.

6. G.S. 19A-21.

7. G.S. 19A-37.

8. G.S. 19A-26.

9. *Id.*

Operations Regulated under the North Carolina Animal Welfare Act

Operation	Definition
Animal shelter	A facility used to house or contain seized, stray, homeless, quarantined, abandoned, or unwanted animals. The term includes shelters under contract with, owned, operated, or maintained by either • a local government (i.e., municipality or county) or • a nonprofit organization devoted to the welfare, protection, rehabilitation, or humane treatment of animals.
Boarding kennel	A facility or establishment that regularly offers to the public, for a fee, the service of boarding dogs or cats or both. The facility may offer other services, such as grooming.
Dealer	Any person who sells, exchanges, or donates animals to any of the following: • other dealers • pet shops • research facilities The term does not include persons who breed and raise on their own premises no more than the offspring of five female dogs or cats per year (unless the animals are bred and raised specifically for research purposes).
Pet shop	A person or establishment that acquires for resale animals bred by others and sells or trades them to the general public at retail or wholesale.

dealers) must obtain licenses.[10] Certificates of registration required for animal shelters are free, whereas the licenses required for the other operations cost fifty dollars each year.[11]

The law allows the Animal Welfare Section to deny an application for a license or certificate of registration and to suspend or revoke a license or certificate for various reasons: for example, willfully disregarding state law or making substantial misrepresentations or false promises in connection with the business to be licensed.[12] Appeals of such decisions are handled as contested cases through the Office of Administrative Hearings.[13]

To obtain a license or certificate of registration, facilities must meet the regulations' detailed requirements for the housing and care of animals.[14] Facilities must, for example, use a specified formula to calculate the minimum square footage of floor space for the primary enclosures for cats and dogs,[15] and they

10. G.S. 19A-27 (pet shop); G.S. 19A-28 (public auctions and boarding kennels); G.S. 19A-29 (dealers).

11. G.S. 19A-26 through 29.

12. G.S. 19A-30 (outlining various grounds for denial, suspension, and revocation and detailing the process for doing so); G.S. 19A-35 (allowing revocation of license or registration for failure to adequately house, feed, and water animals).

13. G.S. 19A-30; G.S. 19A-32.

14. N.C. Admin. Code tit. 2, ch.52J, §§ .0201–.0210 (hereinafter N.C.A.C.).

15. 2 N.C.A.C. 52J .0204(d).

must ensure that enclosures and exercise areas are cleaned at least twice a day.[16] Regulations also govern the transportation of animals.[17]

Euthanasia

In 2005 the General Assembly directed the Board of Agriculture to adopt regulations governing the methods of euthanasia used in animal shelters.[18] As discussed in Chapter 3, page 57, the board considered final regulations in February of 2008, but at the time of publication the rules had not been finalized. The draft rules address two primary subjects: (1) what methods of euthanasia may be used at shelters and (2) who is allowed to euthanize animals at shelters.

With respect to euthanasia methods, state law already provides that animals impounded for violations of the state's rabies control laws may be euthanized by any method approved by one of three national organizations: the American Veterinary Medical Association, the Humane Society of the United States, or the American Humane Association.[19] The draft rules do not change this general guideline in that they also allow the use of any method approved by one of those organizations.[20] (A full list of methods approved by these three organizations is included as Appendix A.) The draft rules include specific requirements related to the two most frequently used euthanasia methods: injection and delivery of carbon monoxide gas in an enclosed chamber. Assuming the rules are finalized in their current form, shelters would still be permitted to use both of these methods, but they would need to comply with significant restrictions on the use of carbon monoxide gas.[21] The draft rules require that

- only commercially compressed, bottled gas be used,
- the gas be delivered in a commercially manufactured chamber,
- animals be individually separated within the chamber,
- animals placed inside the chamber be of the same species,
- live animals not be placed into a chamber with dead animals,
- the chamber rapidly achieve a minimum 6 percent uniform concentration of carbon monoxide,

---

16. 2 N.C.A.C. 52J .0207(a).
17. 2 N.C.A.C. 52J .0301–.0304.
18. S.L. 2005-276, sec. 11.5 (amending G.S. 19A-24(5).
19. G.S. 130A-192.
20. 2 N.C.A.C. 52J .0401.
21. 2 N.C.A.C. 52J .0600). Because the rules have not yet been finalized, citations to the administrative code in this section are somewhat premature. Nevertheless, the anticipated citations are included to assist readres with future research. Before relying upon the citation, it would be wise to verify the actual text of the rule once it has been finalized. The administrative code is available online at http://reports.oah.state.us/ncac.asp.

- death occur within five minutes of beginning the administration of the gas, and
- animals remain in the chamber with the gas for a minimum of twenty minutes.

In addition, the draft rules prohibit the use of gas for euthanizing certain types of animals, including those that are very young or old, pregnant, ill or injured. The draft rules also impose detailed requirements for the construction, ventilation, and maintenance of the chamber.[22] In "extraordinary circumstances," the draft rules allow shelter staff to euthanize an animal by shooting it or using another "extreme" method approved by one of the three national organizations. The rules explain that "an extraordinary circumstance or situation includes but is not limited to a situation which is offsite from the shelter, in which an animal poses a significant or immediate risk to life, limb or public health and in which no alternative, less extreme measure of euthanasia is feasible. It also includes circumstances or situations in which it would be inhumane to transport an animal to another location to perform euthanasia."[23]

With respect to who is authorized to perform euthanasia at a shelter, the draft rules establish a new category of animal control professional: a *certified euthanasia technician*. The term is defined as "a person employed by a certified facility who has been instructed in the proper methods of humane euthanasia, security, and record keeping and possesses other skills as deemed necessary by the Animal Welfare Section."[24] These technicians are required to complete specific training and pass an examination before being certified for up to five years.[25] Under certain circumstances, the Department has the authority to discipline technicians, which could include suspension or revocation of certification.[26] Other professionals authorized to perform euthanasia include a probationary euthanasia technician (in some circumstances), a veterinarian licensed in North Carolina, and a registered veterinary technician working under a veterinarian's supervision.[27]

---

22. 2 N.C.A.C. 52J .0605–.0608.
23. 2 N.C.A.C. 52J .0700.
24. 2 N.C.A.C. 52J .0403.
25. 2 N.C.A.C. 52J .0404–.0407.
26. 2 N.C.A.C. 52J .0419.
27. 2 N.C.A.C. 52J .0402; *see also* 2 N.C.A.C. 52J .0409 (requiring a probationary euthanasia technician to be under the supervision of someone authorized to perform euthanasia).

Penalties

State law provides for both civil and criminal penalties. It is a Class 3 misdemeanor to operate a pet shop, kennel, or public auction without a license and a Class 2 misdemeanor to act as a dealer without a license.[28] It is also a Class 3 misdemeanor for a person who is licensed or registered pursuant to the state law to fail to adequately house, feed, and water animals in his or her care.[29] When a person is charged with either (1) dealing without a license or (2) failing to adequately house, feed, and water animals, the Animal Welfare Section is authorized to immediately seize and impound the animals. If the person charged is subsequently convicted, the state may provide for the sale or euthanasia of the animals.[30]

The director of the Animal Welfare Section has the authority to impose civil monetary penalties against any person who violates the state animal laws or regulations.[31] The proceeds of penalties collected under this law are placed in a fund dedicated to the public schools.[32]

Federal Law

The federal Animal Welfare Act (AWA), which is administered by the Animal and Plant Health Inspection Service (APHIS)[33] of the U.S. Department of Agriculture (USDA), regulates several different categories of animal operations. It focuses on operations that use animals for research, sell animals to the public, transport animals, and use animals in exhibits.[34] There is obviously some overlap in jurisdiction between the federal and state laws, but there are also important distinctions.

The AWA requires some operations, specifically dealers and exhibitors, to obtain licenses.[35] Other regulated operations (research facilities, handlers, transporters, and unlicensed exhibitors) must register with the USDA.[36] The law requires regulated operations to maintain certain records and imposes detailed

---

28. G.S. 19A-33 to 34.
29. G.S. 19A-35.
30. G.S. 19A-34.
31. G.S. 19A-40.
32. *See* Article 31A of G.S. Ch. 115C. When a civil fine is collected by a state agency, the agency is allowed to retain a small amount of the proceeds to cover the costs incurred in collecting the fine. That amount may not exceed 20 percent of the total amount collected. G.S. 115C-457.2.
33. www.aphis.usda.gov/animal_welfare/index.shtml
34. 7 U.S.C. §§ 2131–2159.
35. 7 U.S.C. § § 2133–2134.
36. 7 U.S.C. § 2136.

requirements regarding the handling, care, treatment, and transportation of animals.[37]

One provision of the federal law applies directly to local government animal shelters. It requires shelters to hold all dogs and cats for a minimum of five days before selling them to a dealer.[38] This holding requirement also applies to private entities (such as humane societies) that are under contract to a state, county or city, or research facility licensed by the USDA.

The law authorizes the USDA to conduct investigations and inspections, impose civil money penalties, and apply for injunctions.[39] Licensees may also be subject to criminal penalties.[40]

Operations Regulated under the Federal Animal Welfare Act[41]

Operation	Definition
Carriers	The operator of any airline, railroad, motor carrier, shipping line, or other enterprise engaged in the business of transporting any animals for hire.
Dealers	Any person who, for compensation or profit, delivers for transportation or transports (except as a carrier), or buys, sells, or negotiates the purchase or sale of (1) any animal (alive or dead) for research, teaching, exhibition, or use as a pet or (2) any dog for hunting, security, or breeding purposes. The term does not include retail pet stores (except a pet store that sells animals to research facilities, exhibitors, or dealers) and private persons who earn no more than $500 a year selling animals other than dogs, cats, and wild animals.
Exhibitors	Any person exhibiting any animal involved in commerce (e.g., one purchased or exhibited to the public for compensation). Includes carnivals, circuses, and zoos, whether operated for profit or not. Excludes retail pet stores, state and county fairs, livestock shows, rodeos, purebred dog and cat shows, and other exhibitions determined by the Department of Agriculture.
Intermediate handlers	Any person engaged in any business who receives custody of animals in connection with their transportation in commerce.
Research facilities	Any school (except an elementary or secondary school), institution, organization, or person that uses or intends to use live animals in research, tests, or experiments. The operation must also either (1) purchase or transport live animals in commerce or (2) receive federal funds for the purpose of supporting the research. Even if a facility meets the definition, the Department of Agriculture has the authority to exempt certain facilities under some circumstances.

37. 7 U.S.C. §§ 2140, 2143, 2144; 9 C.F.R. Parts 2 and 3.
38. 7 U.S.C. § 2158(a).
39. 7 U.S.C. §§ 2146–2147, 2149. 2159.
40. 7 U.S.C. § 2149(c)
41. 7 U.S.C. § 2132.

Relevant Statutes

Article 3 of Chapter 19A
Animal Welfare Act.

§ 19A-20. Title of Article.
This Article may be cited as the Animal Welfare Act.

§ 19A-21. Purposes.
The purposes of this Article are (i) to protect the owners of dogs and cats from the theft of such pets; (ii) to prevent the sale or use of stolen pets; (iii) to insure that animals, as items of commerce, are provided humane care and treatment by regulating the transportation, sale, purchase, housing, care, handling and treatment of such animals by persons or organizations engaged in transporting, buying, or selling them for such use; (iv) to insure that animals confined in pet shops, kennels, animal shelters and auction markets are provided humane care and treatment; (v) to prohibit the sale, trade or adoption of those animals which show physical signs of infection, communicable disease, or congenital abnormalities, unless veterinary care is assured subsequent to sale, trade or adoption.

§ 19A-22. Animal Welfare Section in Animal Health Division of Department of Agriculture and Consumer Services created; Director.
There is hereby created within the Animal Health Division of the North Carolina Department of Agriculture and Consumer Services, a new section thereof, to be known as the Animal Welfare Section of said division.

The Commissioner of Agriculture is hereby authorized to appoint a Director of said section whose duties and authority shall be determined by the Commissioner subject to the approval of the Board of Agriculture and subject to the provisions of this Article.

§ 19A-23. Definitions.
For the purposes of this Article, the following terms, when used in the Article or the rules or orders made pursuant thereto, shall be construed respectively to mean:

(1) "Adequate feed" means the provision at suitable intervals, not to exceed 24 hours, of a quantity of wholesome foodstuff suitable for the species and age, sufficient to maintain a reasonable level of nutrition in each animal. Such foodstuff shall be served in a sanitized receptacle, dish, or container.

(2) "Adequate water" means a constant access to a supply of clean, fresh, potable water provided in a sanitary manner or provided at suitable intervals for the species and not to exceed 24 hours at any interval.

(3) "Ambient temperature" means the temperature surrounding the animal.

(4) "Animal" means any domestic dog (Canis familiaris), or domestic cat (Felis domestica).

(5) "Animal shelter" means a facility which is used to house or contain seized, stray, homeless, quarantined, abandoned or unwanted animals and which is under contract with, owned, operated, or maintained by a county, city, town, or other municipality, or by a duly incorporated humane society, animal welfare society, society for the prevention of cruelty to animals, or other nonprofit organization devoted to the welfare, protection, rehabilitation, or humane treatment of animals.

(5a) "Boarding kennel" means a facility or establishment which regularly offers to the public the service of boarding dogs or cats or both for a fee. Such a facility or establishment may, in addition to providing shelter, food and water, offer grooming or other services for dogs and/or cats.

(6) "Commissioner" means the Commissioner of Agriculture of the State of North Carolina.

(7) "Dealer" means any person who sells, exchanges, or donates, or offers to sell, exchange, or donate animals to another dealer, pet shop, or research facility; provided, however, that an individual who breeds and raises on his own premises no more than the offspring of five canine or feline females per year, unless bred and raised specifically for research purposes shall not be considered to be a dealer for the purposes of this Article.

(8) "Director" means the Director of the Animal Welfare Section of the Animal Health Division of the Department of Agriculture and Consumer Services.

(9) "Euthanasia" means the humane destruction of an animal accomplished by a method that involves rapid unconsciousness and immediate death or by a method that involves anesthesia, produced by an agent which causes painless loss of consciousness, and death during such loss of consciousness.

(10) "Housing facility" means any room, building, or area used to contain a primary enclosure or enclosures.

(11) "Person" means any individual, partnership, firm, joint-stock company, corporation, association, trust, estate, or other legal entity.
(12) "Pet shop" means a person or establishment that acquires for the purposes of resale animals bred by others whether as owner, agent, or on consignment, and that sells, trades or offers to sell or trade such animals to the general public at retail or wholesale.
(13) "Primary enclosure" means any structure used to immediately restrict an animal or animals to a limited amount of space, such as a room, pen, cage compartment or hutch.
(14) "Public auction" means any place or location where dogs or cats are sold at auction to the highest bidder regardless of whether such dogs or cats are offered as individuals, as a group, or by weight.
(15) "Research facility" means any place, laboratory, or institution at which scientific tests, experiments, or investigations involving the use of living animals are carried out, conducted, or attempted.
(16) "Sanitize" means to make physically clean and to remove and destroy to a practical minimum, agents injurious to health.

§ 19A-24. Powers of Board of Agriculture.

The Board of Agriculture shall:

(1) Establish standards for the care of animals at animal shelters, boarding kennels, pet shops, and public auctions. A boarding kennel that offers dog day care services and has a ratio of dogs to employees or supervisors, or both employees and supervisors, of not more than 10 to one, shall not as to such services be subject to any regulations that restrict the number of dogs that are permitted within any primary enclosure.
(2) Prescribe the manner in which animals may be transported to and from registered or licensed premises.
(3) Require licensees and holders of certificates to keep records of the purchase and sale of animals and to identify animals at their establishments.
(4) Adopt rules to implement this Article, including federal regulations promulgated under Title 7, Chapter 54, of the United States Code.
(5) Adopt rules on the euthanasia of animals in the possession or custody of any person required to obtain a certificate of registration under this Article. An animal shall only be put to death by a method and delivery of method approved by the American Veterinary Medical Association, the Humane Society of the United States, or the American Humane Association. The Department shall establish

rules for the euthanasia process using any one or combination of methods and standards prescribed by the three aforementioned organizations. The rules shall address the equipment, the process, and the separation of animals, in addition to the animals' age and condition. If the gas method of euthanasia is approved, rules shall require (i) that only commercially compressed carbon monoxide gas is approved for use, and (ii) that the gas must be delivered in a commercially manufactured chamber that allows for the individual separation of animals. Rules shall also mandate training for any person who participates in the euthanasia process.

§ 19A-25. Employees; investigations; right of entry.

For the enforcement of the provisions of this Article, the Director is authorized, subject to the approval of the Commissioner to appoint employees as are necessary in order to carry out and enforce the provisions of this Article, and to assign them interchangeably with other employees of the Animal Health Division. The Director shall cause the investigation of all reports of violations of the provisions of this Article, and the rules adopted pursuant to the provisions hereof; provided further, that if any person shall deny the Director or his representative admittance to his property, either person shall be entitled to secure from any superior court judge a court order granting such admittance.

§ 19A-26. Certificate of registration required for animal shelter.

No person shall operate an animal shelter unless a certificate of registration for such animal shelter shall have been granted by the Director. Application for such certificate shall be made in the manner provided by the Director. No fee shall be required for such application or certificate. Certificates of registration shall be valid for a period of one year or until suspended or revoked and may be renewed for like periods upon application in the manner provided.

§ 19A-27. License required for operation of pet shop.

No person shall operate a pet shop unless a license to operate such establishment shall have been granted by the Director. Application for such license shall be made in the manner provided by the Director. The license shall be for the fiscal year and the license fee shall be fifty dollars ($50.00) for each license period or part thereof beginning with the first day of the fiscal year.

§ 19A-28. License required for public auction or boarding kennel.

No person shall operate a public auction or a boarding kennel unless a license to operate such establishment shall have been granted by the Director.

Application for such license shall be made in the manner provided by the Director. The license period shall be the fiscal year and the license fee shall be fifty dollars ($50.00) for each license period or part thereof beginning with the first day of the fiscal year.

§ 19A-29. License required for dealer.
No person shall be a dealer unless a license to deal shall have been granted by the Director to such person. Application for such license shall be in the manner provided by the Director. The license period shall be the fiscal year and the license fee shall be fifty dollars ($50.00) for each license period or part thereof, beginning with the first day of the fiscal year.

§ 19A-30. Refusal, suspension or revocation of certificate or license.
The Director may refuse to issue or renew or may suspend or revoke a certificate of registration for any animal shelter or a license for any public auction, kennel, pet shop, or dealer, if after an impartial investigation as provided in this Article he determines that any one or more of the following grounds apply:

(1) Material misstatement in the application for the original certificate of registration or license or in the application for any renewal under this Article;

(2) Willful disregard or violation of this Article or any rules issued pursuant thereto;

(3) Failure to provide adequate housing facilities and/or primary enclosures for the purposes of this Article, or if the feeding, watering, sanitizing and housing practices at the animal shelter, public auction, pet shop, or kennel are not consistent with the intent of this Article or the rules adopted under this Article;

(4) Allowing one's license under this Article to be used by an unlicensed person;

(5) Conviction of any crime an essential element of which is misstatement, fraud, or dishonesty, or conviction of any felony;

(6) Making substantial misrepresentations or false promises of a character likely to influence, persuade, or induce in connection with the business of a public auction, commercial kennel, pet shop, or dealer;

(7) Pursuing a continued course of misrepresentation of or making false promises through advertising, salesmen, agents, or otherwise in connection with the business to be licensed;

(8) Failure to possess the necessary qualifications or to meet the requirements of this Article for the issuance or holding of a certificate of registration or license.

The Director shall, before refusing to issue or renew and before suspension or revocation of a certificate of registration or a license, give to the applicant or holder thereof a written notice containing a statement indicating in what respects the applicant or holder has failed to satisfy the requirements for the holding of a certificate of registration or a license. If a certificate of registration or a license is suspended or revoked under the provisions hereof, the holder shall have five days from such suspension or revocation to surrender all certificates of registration or licenses issued thereunder to the Director or his authorized representative.

A person to whom a certificate of registration or a license is denied, suspended, or revoked by the Director may contest the action by filing a petition under G.S. 150B-23 within five days after the denial, suspension, or revocation.

Any licensee whose license is revoked under the provisions of this Article shall not be eligible to apply for a new license hereunder until one year has elapsed from the date of the order revoking said license or if an appeal is taken from said order of revocation, one year from the date of the order or final judgment sustaining said revocation. Any person who has been an officer, agent, or employee of a licensee whose license has been revoked or suspended and who is responsible for or participated in the violation upon which the order of suspension or revocation was based, shall not be licensed within the period during which the order of suspension or revocation is in effect.

§ 19A-31. License not transferable; change in management, etc., of business or operation.

A license is not transferable. When there is a transfer of ownership, management, or operation of a business of a licensee hereunder, the new owner, manager, or operator, as the case may be, whether it be an individual, firm, partnership, corporation, or other entity shall have 10 days from such sale or transfer to secure a new license from the Director to operate said business. A licensee shall promptly notify the Director of any change in the name, address, management, or substantial control of his business or operation.

§ 19A-32. Procedure for review of Director's decisions.

A denial, suspension, or revocation of a certificate or license under this Article shall be made in accordance with Chapter 150B of the General Statutes.

§ 19A-33. Penalty for operation of pet shop, kennel or auction without license.
Operation of a pet shop, kennel, or public auction without a currently valid license shall constitute a Class 3 misdemeanor subject only to a penalty of not less than five dollars ($5.00) nor more than twenty-five dollars ($25.00), and each day of operation shall constitute a separate offense.

§ 19A-34. Penalty for acting as dealer without license; disposition of animals in custody of unlicensed dealer.
Acting as a dealer in animals as defined in this Article without a currently valid dealer's license shall constitute a Class 2 misdemeanor. Continued illegal operation after conviction shall constitute a separate offense. Animals found in possession or custody of an unlicensed dealer shall be subject to immediate seizure and impoundment and upon conviction of such unlicensed dealer shall become subject to sale or euthanasia in the discretion of the Director.

§ 19A-35. Penalty for failure to adequately care for animals; disposition of animals.
Failure of any person licensed or registered under this Article to adequately house, feed, and water animals in his possession or custody shall constitute a Class 3 misdemeanor, and such person shall be subject to a fine of not less than five dollars ($5.00) per animal or more than a total of one thousand dollars ($1,000). Such animals shall be subject to seizure and impoundment and upon conviction may be sold or euthanized at the discretion of the Director and such failure shall also constitute grounds for revocation of license after public hearing.

§ 19A-36. Penalty for violation of Article by dog warden.
Violation of any provision of this Article which relates to the seizing, impoundment, and custody of an animal by a dog warden shall constitute a Class 3 misdemeanor and the person convicted thereof shall be subject to a fine of not less than fifty dollars ($50.00) and not more than one hundred dollars ($100.00), and each animal handled in violation shall constitute a separate offense.

§ 19A-37. Application of Article.
This Article shall not apply to a place or establishment which is operated under the immediate supervision of a duly licensed veterinarian as a hospital where animals are harbored, boarded, and cared for incidental to the treatment, prevention, or alleviation of disease processes during the routine practice of the profession of veterinary medicine. This Article shall not apply to any dealer, pet shop, public auction, commercial kennel or research facility during the period

such dealer or research facility is in the possession of a valid license or registration granted by the Secretary of Agriculture pursuant to Title 7, Chapter 54, of the United States Code. This Article shall not apply to any individual who occasionally boards an animal on a noncommercial basis, although such individual may receive nominal sums to cover the cost of such boarding.

§ 19A-38. Use of license fees.
All license fees collected shall be used in enforcing and administering this Article.

§ 19A-39. Article inapplicable to establishments for training hunting dogs.
Nothing in this Article shall apply to those kennels or establishments operated primarily for the purpose of boarding or training hunting dogs.

§ 19A-40. Civil Penalties.
The Director may assess a civil penalty of not more than five thousand dollars ($5,000) against any person who violates a provision of this Article or any rule promulgated thereunder. In determining the amount of the penalty, the Director shall consider the degree and extent of harm caused by the violation. The clear proceeds of civil penalties assessed pursuant to this section shall be remitted to the Civil Penalty and Forfeiture Fund in accordance with G.S. 115C-457.2.

§ 19A-41. Legal representation by the Attorney General.
It shall be the duty of the Attorney General to represent the Commissioner of Agriculture and the Department of Agriculture and Consumer Services, or to designate some member of his staff to represent the Commissioner and the Department, in all actions or proceedings in connection with this Article.

---

§ 19A-65. Annual Report Required From Every Animal Shelter in Receipt of State or Local Funding.
Every county or city animal shelter, or animal shelter operated under contract with a county or city or otherwise in receipt of State or local funding shall prepare an annual report setting forth the numbers, by species, of animals received into the shelter, the number adopted out, the number returned to owner, and the number destroyed. The report shall also contain the total operating expenses of the shelter and the cost per animal handled. The report shall be filed with the Department of Health and Human Services by August 1 of each year.

§ 153A-442. Animal shelters.
A county may establish, equip, operate, and maintain an animal shelter or may contribute to the support of an animal shelter, and for these purposes may appropriate funds not otherwise limited as to use by law. The animal shelters shall meet the same standards as animal shelters regulated by the Department of Agriculture pursuant to its authority under Chapter 19A of the General Statutes.

§ 153A-449. Contracts with private entities.
A county may contract with and appropriate money to any person, association, or corporation, in order to carry out any public purpose that the county is authorized by law to engage in.

§ 160A-20.1. Contracts with private entities.
A city may contract with and appropriate money to any person, association, or corporation, in order to carry out any public purpose that the city is authorized by law to engage in.

§ 160A-493. Animal shelters.
A city may establish, equip, operate, and maintain an animal shelter or may contribute to the support of an animal shelter, and for these purposes may appropriate funds not otherwise limited as to use by law. The animal shelters shall meet the same standards as animal shelters regulated by the Department of Agriculture pursuant to its authority under Chapter 19A of the General Statutes.

Chapter 8

Spay/Neuter Program

According to the American Veterinary Medical Association (AVMA), the number of unwanted pets in this country is "a primary welfare concern of our society."[1] The AVMA, animal welfare organizations, and others have long encouraged state and local governments to take a more active role in controlling the pet populations.[2] In 2000 the General Assembly adopted Article 5 of North Carolina General Statute 19A (hereinafter G.S.) establishing the Spay/Neuter Program. The program provides funding to local governments that offer spay and neuter services at a reduced cost to low-income persons. The law also directed the state to establish an education program promoting the benefits of spaying and neutering pets.

---

1. American Veterinary Medical Association, Policy Statement: Dog and Cat Population Control (June 2005), www.avma.org/issues/policy/animal_welfare/population_control.asp (last visited June 15, 2007).

2. *Id.* (recommending that state and local government provide more funding for animal control, prohibit the sale or adoption of intact animals by animal control and humane organizations, and promote sterilization); *see also* Humane Society of the United States, Solving the Pet Overpopulation Problem (Oct. 12, 2006), www.hsus.org/pets/issues_affecting_our_pets/pet_overpopulation_and_ownership_statistics/solving_the_pet_overpopulation_problem.html (last visited June 15, 2007) (advocating community-based solutions, including mandatory sterilization of shelter adoptees, low-cost spay and neuter programs, and differential licensing); Jean McNeil and Elisabeth Constandy, "Addressing the Problem of Pet Overpopulation: The Experience of New Hanover County Animal Control Services," *Journal of Public Health Management and Practice* 12 (2006): 452. *But see* North Carolina Responsible Animal Owners' Alliance, "The Myth of Over Population" (Feb. 2006), www.ncraoa.com/articles/canine/OverPopulationMyth.html (last visited June 15, 2007) (arguing that irresponsible owners are the primary reason for the high numbers of animals euthanized in shelters).

The Spay/Neuter Program is administered by the Veterinary Public Health Program in the North Carolina Department of Health and Human Services (DHHS).[3] There are three funding sources for the program.

- twenty cents of the fee collected for each rabies tag purchased from the state[4]
- ten dollars of the additional fee imposed for a special "Animal Lovers" license plate[5]
- other funds, such as appropriations and private contributions

When the law was initially enacted in 2000, it included a funding stream generated by a surcharge of fifty cents paid by individuals choosing a special "I Care" rabies tag. This special tag option was eliminated in 2007 and replaced with the mandatory twenty-cent surcharge on every rabies tag sold by the state.[6] Despite this mandatory surcharge, this change in the law does not establish a guaranteed funding stream for the account because not every rabies tag issued in the state is sold by the state. Many veterinarians and clinics purchase tags from private vendors.

According to the statute, funds in the Spay/Neuter Program account must be allocated as follows:

- Up to $47,500 may be used to fund rabies education and prevention programs.
- Twenty percent must be dedicated to the development and implementation of the statewide education program.
- Up to 20 percent may be used by DHHS for the administration of the Spay/Neuter Program.
- The remainder must be distributed to local governments.[7]

---

3. More information about the Spay/Neuter Program is available at www.epi.state.nc.us/epi/vet/index.html (last visited Jan. 29, 2008).

4. N.C. Gen. Stat. 19A-62(a)(1) (hereinafter G.S.); *see also* G.S. 130A-190(b)(4) (specifying the rabies tag fee). Note that these two statutes were amended by S.L. 2007-487. As of June 2008, the Commission for Public Health regulations governing the rabies tag fee had not yet been updated to reflect the changes in the statutory authority. *See* N.C. Admin. Code tit. 10A, ch. 41G, § .0102 (hereinafter N.C.A.C.) (regulation governing fees for rabies tags, links, and rivets).

5. G.S. 19A-62(a)(2); *see* G.S. 20-79.4 (authorizing the Division of Motor Vehicles to issue "animal lovers" license plates). Information about ordering the specialized plates is available through the North Carolina Division of Motor Vehicles, https://edmv-sp.dot.state.nc.us/sp/SpecialPlatesPortal.html (last visited Mar. 3, 2008).

6. S.L. 2007-487.

7. G.S. 19A-62(b).

The law authorizes DHHS to provide funding for a pilot program to help a single county establish a new animal control program.[8] According to the state public health veterinarian, the state invited applications for a pilot program in 2001 but no counties expressed an interest.[9]

Local Government Funding

As noted above, one purpose of the Spay/Neuter Program is to encourage local governments to establish and maintain systems to help low-income persons afford to spay or neuter their cats and dogs. A *low-income person* is defined as anyone who qualifies for a public assistance program administered by DHHS pursuant to G. S.108A (e.g., Medicaid or Work First).[10] It is important to note that the person must only qualify for a Chapter 108A program; it is not required that he or she actually receive benefits under the program.

A local government may be eligible for funding if it offers one or more of the following services to low-income persons:

- a spay/neuter clinic operated by the county or city,
- a spay/neuter clinic operated by a private organization under contract or other arrangement with the county or city,[11]
- a program or procedure offering low-income pet owners a discount on spaying or neutering services in the community (such as a voucher program), or
- a procedure providing discounted spay and neuter services for animals adopted from an animal shelter operated by or under contract with the county or city.[12]

The service provided (e.g., clinic or discount program) must be available on a year-round basis. In 2007 the General Assembly placed a new condition on receipt of funds from the program. The local government must require animal owners who present their pets for a subsidized spay or neuter service to either

---

8. S.L. 2000-163, sec. 6. Under the law, funding for the pilot program was capped at the lesser of either $50,000 or 50 percent of the funds received from the sale of special license plates.

9. E-mail correspondence from Dr. Carl Williams, state public health veterinarian (Oct. 3, 2007).

10. G.S. 19A-63(b).

11. The law does not specifically state that these clinics must provide spay and neuter services at a reduced cost, but such a requirement could easily be inferred from the statutory context.

12. G.S. 19A-63(a).

(1) show proof of rabies vaccination or (2) have the animal vaccinated at the time of the spay or neuter procedure.[13]

Eligible cities and counties may be reimbursed for the direct costs of the surgical procedure, including anesthesia, medication, and veterinary services. Capital expenditures and administrative costs are not eligible for reimbursement.[14] Applications for reimbursement must be submitted to DHHS quarterly.[15] If there are not enough funds available to reimburse all applicants, the statute specifies how the funds should be allocated. Specifically, the act requires that 50 percent of the available funds be distributed to the most economically distressed counties.[16] The amount of funding available to each county also depends on the proportion of dogs and cats vaccinated.[17] In 2006 about twenty jurisdictions and organizations received funding from the program; reimbursements ranged from over $18,000 for Robeson County to just over $1,000 for Randolph County and the City of Laurinburg.[18]

A city or county that receives funding from the program is required to report certain data related to shelter operations to DHHS. Specifically, jurisdictions must report the number of animals received into the shelter (by species) as well as the dispositions of animals (i.e., how many were adopted out, returned to their owner, or destroyed). The report must also state the total operating expenses of the shelter and the cost per animal handled. In 2005 the cost per animal handled ranged from about $25 to over $400.[19]

---

13. G.S. 19A-62(b)(4) (as amended by S.L. 2007-487).

14. G.S. 19A-64(a).

15. Applications and other information related to the application process are available from the North Carolina Department of Health and Human Services (hereinafter DHHS), Veterinary Public Health Program, www.rabies.ncdhhs.gov/epi/vet/index.html (last visited Oct. 12, 2007).

16. The North Carolina Department of Commerce evaluates the financial conditions of each county on an annual basis and places it into one of five tiers, with tier one being the most economically distressed and tier five being the least distressed. G.S. 143B-437.08. The Spay/Neuter Program gives reimbursement priority to counties in tiers one, two, and three.

17. G.S. 19A-64(c)(2)–(3).

18. Spay/Neuter Program, Division of Public Health, DHHS Services, S/N Reimbursement, by City/County (May 14, 2007), www.epi.state.nc.us/epi/vet/pdf/SNCityCountyReqbyQtr.pdf (last visited June 20, 2007).

19. Animal Shelter Reporting Data (Sept. 4, 2007), www.epi.state.nc.us/epi/vet/pdf/snreport2006.pdf.

Relevant Statutes

Article 5 of Chapter 19A
Spay/Neuter Program.

§ 19A-60. Legislative findings.
The General Assembly finds that the uncontrolled breeding of cats and dogs in the State has led to unacceptable numbers of unwanted dogs, puppies and cats and kittens. These unwanted animals become strays and constitute a public nuisance and a public health hazard. The animals themselves suffer privation and death, are impounded, and most are destroyed at great expense to local governments. It is the intention of the General Assembly to provide a voluntary means of funding a spay/neuter program to provide financial assistance to local governments offering low-income persons reduced-cost spay/neuter services for their dogs and cats and to provide a statewide education program on the benefits of spaying and neutering pets.

§ 19A-61. Spay/Neuter Program established.
There is established in the Department of Health and Human Services a statewide program to foster the spaying and neutering of dogs and cats for the purpose of reducing the population of unwanted animals in the State. The program shall consist of the following components:

(1) Education Program. – The Department shall establish a statewide program to educate the public about the benefits of having cats and dogs spayed and neutered. The Department may work cooperatively on the program with the North Carolina School of Veterinary Medicine, other State agencies and departments, county and city health departments and animal control agencies, and statewide and local humane organizations. The Department may employ outside consultants to assist with the education program.

(2) Local Spay/Neuter Assistance Program. – The Department shall administer the Spay/Neuter Account established in G.S. 19A-62. Monies deposited in the account shall be available to reimburse eligible counties and cities for the direct costs of spay/neuter surgeries for cats and dogs made available to low-income persons.

§ 19A-62. Spay/Neuter Account established.

(a) Creation. – The Spay/Neuter Account is established as a nonreverting special revenue account in the Department of Health and Human Services. The Account consists of the following:

(1) The portion of the fee imposed under G.S. 130A-190(b)(4) for obtaining a rabies vaccination tag from the Department of Health and Human Services.
(2) Ten dollars ($10.00) of the additional fee imposed by G.S. 20-79.7 for an Animal Lovers special license plate.
(3) Any other funds available from appropriations by the General Assembly or from contributions and grants from public or private sources.

(b) Use. – The revenue in the Account shall be used by the Department of Health and Human Services as follows:
(1) If the revenue generated by the portion of the fee imposed under G.S. 130A-190(b)(3) is less than forty-seven thousand five hundred dollars ($47,500) for the fiscal year, then funds up to the difference between forty-seven thousand five hundred dollars ($47,500) and the amount of revenue generated may be used from this Account to fund rabies education and prevention programs.
(2) Twenty percent (20%) shall be used to develop and implement the statewide education program component of the Spay/Neuter Program established in G.S. 19A-61(a).
(3) Up to twenty percent (20%) of the money in the Account may be used to defray the costs of administering the Spay/Neuter Program established in this Article.
(4) Funds remaining after deductions for the education program and administrative expenses shall be distributed quarterly to eligible counties and cities seeking reimbursement for reduced-cost spay/neuter surgeries performed during the previous year. A county or city is ineligible to receive funds under this subdivision unless it requires the owner to show proof of rabies vaccination at the time of the procedure or, if none, require vaccination at the time of the procedure.

§ 19A-63. Eligibility for distributions from Spay/Neuter Account.

(a) A county or city is eligible for reimbursement from the Spay/Neuter Account if it meets the following condition:
(1) The county or city offers one or more of the following programs to low-income persons on a year-round basis for the purpose of reducing the cost of spaying and neutering procedures for dogs and cats:
a. A spay/neuter clinic operated by the county or city.

b. A spay/neuter clinic operated by a private organization under contract or other arrangement with the county or city.
c. A contract or contracts with one or more veterinarians, whether or not located within the county, to provide reduced-cost spaying and neutering procedures.
d. Subvention of the spaying and neutering costs incurred by low-income pet owners through the use of vouchers or other procedure that provides a discount of the cost of the spaying or neutering procedure fixed by a participating veterinarian or other provider.
e. Subvention of the spaying and neutering costs incurred by persons who adopt a pet from an animal shelter operated by or under contract with the county or city.

(2) Reserved for future codification purposes.

(b) For purposes of this Article, the term "low-income person" shall mean an individual who qualifies for one or more of the programs of public assistance administered by the Department pursuant to Chapter 108A of the General Statutes.

§ 19A-64. Distributions to counties and cities from Spay/Neuter Account.

(a) Reimbursable Costs. – Counties and cities eligible for distributions from the Spay/Neuter Account may receive reimbursement for the direct costs of a spay/neuter surgical procedure for a dog or cat owned by a low-income person meeting the Department's eligibility requirements for spay/neuter services. Reimbursable costs shall include anesthesia, medication, and veterinary services. Counties and cities shall not be reimbursed for the administrative costs of providing reduced-cost spay/neuter services or capital expenditures for facilities and equipment associated with the provision of such services.

(b) Application. – A county or city eligible for reimbursement of spaying and neutering costs from the Spay/Neuter Account shall apply to the Department of Health and Human Services by the last day of January, April, July, and October of each year to receive a distribution from the Account for that quarter. The application shall be submitted in the form required by the Department and shall include an itemized listing of the costs for which reimbursement is sought.

(c) Distribution. – The Department shall make payments from the Spay/Neuter Account to eligible counties and cities who have made timely application for reimbursement within 30 days of the closing date for receipt of applications for that quarter. In the event that total requests for reimbursement exceed the amounts available in the Spay/Neuter Account for distribution, the monies available will be distributed as follows:

(1) Fifty percent (50%) of the monies available in the Spay/Neuter Account shall be reserved for reimbursement for eligible applicants within enterprise tier one, two, and three areas as defined in G.S. 105-129.3. The remaining fifty percent (50%) of the funds shall be used to fund reimbursement requests from eligible applicants in enterprise tier four and five areas as defined in G.S. 105-129.3.

(2) Among the eligible counties and cities in enterprise tier one, two, and three areas, reimbursement shall be made to each eligible county or city in proportion to the number of dogs and cats that have received rabies vaccinations during the preceding fiscal year in that county or city as compared to the number of dogs and cats that have received rabies vaccinations during the preceding fiscal year by all of the eligible applicants in enterprise tier one, two, or three areas.

(3) Among the eligible counties and cities in enterprise tier four and five areas, reimbursement shall be made to each eligible county or city in proportion to the number of dogs and cats that have received rabies vaccinations during the preceding fiscal year in that county or city as compared to the number of dogs and cats that have received rabies vaccinations during the preceding fiscal year by all of the eligible applicants in enterprise tier four and five areas.

(4) Should funds remain available from the fifty percent (50%) of the Spay/Neuter Account designated for enterprise tier one, two, or three areas after reimbursement of all claims by eligible applicants in those areas, the remaining funds shall be made available to reimburse eligible applicants in enterprise tier four and five areas.

---

§ 130A-190. Rabies vaccination tags.

(a) Issuance – A licensed veterinarian or a certified rabies vaccinator who administers rabies vaccine to a dog or cat shall issue a rabies vaccination tag to the owner of the animal. The rabies vaccination tag shall show the year issued, a vaccination number, the words "North Carolina" or the initials "N.C." and the words "rabies vaccine." Dogs and cats shall wear rabies vaccination tags at all times. However, cats may be exempted from wearing the tags by local ordinance.

(b) Fee. – Rabies vaccination tags, links, and rivets may be obtained from the Department. The Secretary is authorized to establish by rule a fee for the rabies tags, links and, rivets in accordance with this subsection. The fee for each tag is the sum of the following:

(1) The actual cost of the rabies tag, links, and rivets.
(2) Transportation costs.
(3) Five cents (5¢). This portion of the fee shall be used to fund rabies education and prevention programs.
(4) Twenty cents (20¢). This portion of the fee shall be credited to the Spay/Neuter Account established in G.S. 19A-62 and used to fund statewide spay/neuter programs. This portion of the fee shall not be imposed for tags provided to persons who operate establishments primarily for the purpose of boarding or training hunting dogs or who own and vaccinate 10 or more dogs per year.

Chapter 9

Service Animals

Some individuals with disabilities use service or assistance animals to help them carry out their daily activities. These animals are most often dogs, but other animals, such as monkeys and horses, also serve in this role.[1] People with disabilities who rely on such animals have various legal protections at both the federal and state levels. Members of the public often rely on local animal control officers to help them interpret and apply the laws pertaining to these animals.

Terminology

Two different terms—*service animal* and *assistance animal*—are used in state and federal laws on this topic. *Service animal* is used in the context of the Americans with Disabilities Act of 1990 (ADA), a federal law that affords certain rights to persons with disabilities.[2] The ADA defines a service animal as

> any guide dog, signal dog, or other animal individually trained to do work or perform tasks for the benefit of an individual with a disability, including, but not limited to, guiding individuals with impaired vision, alerting individuals with impaired hearing to intruders or sounds, providing minimal protection or rescue work, pulling a wheelchair, or fetching dropped items.[3]

---

1. Kristina Adams and Stacy Rice, "A Brief Information Resource on Assistance Animals for the Disabled," U.S. Department of Agriculture, Animal Welfare Information Center (April 2004), www.nal.usda.gov/awic/companimals/assist.htm (last visited June 12, 2007).
2. 42 U.S.C. § 12101–12189.
3. 28 C.F.R. § 36.104.

When determining whether an animal meets this definition, courts often evaluate whether it has either (1) been trained to assist a person with a disability or (2) is peculiarly suited to alleviate the problems associated with the person's disability.[4] Although courts often call for some evidence of individual training that sets the service animal apart from other animals, there are no federally mandated training standards.[5]

The term *assistance animal* is used in the context of a North Carolina state law imposing criminal penalties for harming such animals. The term is defined as "an animal that is trained and may be used to assist a 'person with a disability'" and specifically states that it applies to any type of animal, not just dogs. [6]

Federal Law

The ADA is a federal law that, among other things, requires places of public accommodation to "modify policies, practices, or procedures to permit the use of a service animal by an individual with a disability."[7] The U.S. Department of Justice (DOJ) has issued guidance explaining that "businesses and organizations that serve the public must allow people with disabilities to bring their service animals into all areas of the facility where customers are normally allowed to go." [8] A business may ask the person if the animal is a service animal or ask what tasks the animal has been trained to perform, but it must not inquire about the person's disability.[9] It is possible that a place of public accommodation

---

4. *See, e.g.,* Prindable v. Ass'n of Apt. Owners of 2987 Kalakaua, 304 F. Supp. 2d 1245 (D. Haw. 2003) (rejecting plaintiff's claim under the Fair Housing Act that his dog was a service animal); Baugher v. City of Ellensburg, 2007 WL 858627, at *6 (E.D. Wash. March 19, 2007) (rejecting plaintiff's claim under the ADA that her dog was a service animal); Storms v. Fred Meyer Stores, 129 Wash. App. 820 (2005) (recognizing that a plaintiff's dog received some training to assist with her disability and therefore could be considered a service animal under the ADA and state law).

5. *Prindable,* 304 F. Supp. at 1256.

6. N.C. Gen. Stat.§ 14-163.1 (hereinafter G.S.). The term *person with a disability* is defined in G.S. 168A-3.

7. 28 C.F.R. § 36.302(c)(1). The term *place of public accommodation* is defined broadly to include most private businesses whose "operations affect commerce" and fall within one of several categories, including lodging facilities (except certain bed and breakfast establishments), restaurants, theaters, museums, day care centers, and golf courses. 28 C.F.R. § 36.104.

8. U.S. Department of Justice, ADA Business Brief: Service Animals (April 2002) (herinafter ADA Business Brief), www.ada.gov/svcanimb.htm (last visited May 22, 2008).

9. *Id. See also* Grill v. Costco, 312 F. Supp. 2d 1349, 1353 (2004) (holding that businesses may ask questions about the "task or function" of the service animal and whether the animal is a service animal, but not about the person's disability).

could refuse entry to a service animal if it can show that doing so would fundamentally alter the nature of the services it provides.[10] Federal guidance also provides that a place of public accommodation may exclude an animal if "(1) the animal is out of control and the animal's owner does not take effective action to control it (for example, a dog that barks repeatedly during a movie) or (2) the animal poses a direct threat to the health or safety of others."[11] Other federal laws specifically address rights with respect to service animals in (1) most modes of public transportation (e.g., buses, trains),[12] (2) air carriers,[13] and (3) housing.[14]

A person who believes that his or her federal rights related to service animals have been violated may be able to bring a private lawsuit against the business, transportation provider, or housing provider.[15] Grievances against air carriers may be filed directly with the carriers, as federal law requires them to have complaint-resolution procedures in place and to report disability-related complaints to the U.S. Department of Transportation.[16] Consumers may also file complaints related to air carriers with that department.[17]

---

10. *See* PGA Tour, Inc. v. Martin, 532 U.S. 661, 683 (2001) (quoting 42 U.S.C. § 12182(b)(2)(A)(ii); *see also* Lentini v. California Ctr. for the Arts, 370 F.3d 837, 847 (9th Cir. 2004) (rejecting an argument that the presence of a service dog during a concert fundamentally altered the nature of the concerts); Johnson v. Gambrinus Co./Spoetzl Brewery, 116 F.3d 1052, 1059 (1997) (rejecting a brewery's claim that the presence of a service dog fundamentally altered the nature of a brewery tour).

11. ADA Business Brief.

12. 42 U.S.C. § 12142–12144; *see also* 49 C.F.R. § 37.5 (prohibiting discrimination against persons with disabilities in public transportation); 49 C.F.R. § 27.7 (prohibiting discrimination against persons with disabilities in transportation programs receiving federal financial assistance).

13. *See* 14 C.F.R. § 382.55(a)(2) ("Carriers shall permit a service animal to accompany a qualified individual with a disability in any seat in which the person sits, unless the animal obstructs an aisle or other area that must remain unobstructed in order to facilitate an emergency evacuation."); *see also* Air Carrier Access Act of 1986, 49 U.S.C. § 41705; 14 C.F.R. § 382.37 (requiring carrier to offer alternative seat assignments to persons with disabilities accompanied by a service animal).

14. Fair Housing Act, 42 U.S.C. § 3601–3619; *see also* Joint Statement of the Department of Housing and Urban Development and the Department of Justice, Reasonable Accommodations under the Fair Housing Act (May 17, 2004), www.hud.gov/offices/fheo/library/huddojstatement.pdf (last visited June 21, 2007).

15. 42 U.S.C. § 12133 (incorporating by reference 29 U.S.C. § 794(a) and 42 U.S.C. § 12188 (defining the remedies available under the ADA); 42 U.S.C. § 3613(a) (allowing an aggrieved person to bring a civil action under the FHA).

16. 14 C.F.R. § 382.65 (requiring carriers to have a complaint resolution mechanism); 14 C.F.R. § 382.70 (requiring annual reports of all disability-related complaints).

17. 49 U.S.C. § 41705(c) (requiring the Department of Transportation to publish disability-related complaint data and also submit annual reports to Congress regarding

State Law

State law is similar to the federal law in many ways. Like the ADA, North Carolina law protects the access of persons with assistance animals to places that are otherwise open to the general public. Specifically, the law provides that "every person with a disability has the right to be accompanied by a service animal" in the following places:

- public conveyances (e.g., buses, airplanes)
- lodging places (e.g., hotels)
- places of public accommodation, amusement, or resort[18]

A person with a disability also has a right to keep a service animal "on any premises the person leases, rents, or uses."[19]

A person is entitled to these rights under state law if he or she either (1) has a service animal tag issued by the North Carolina Department of Health and Human Services (DHHS) or (2) shows that the animal is being trained or is trained as a service animal.[20] Note that the service animal tag issued by the state is not required. If the disabled person can demonstrate that the animal is trained as a service animal, the state statute, like the federal law, requires the facility to allow the person to be accompanied by the animal.

People who are training a service animal also have rights with respect to access to public places. Trainers may be accompanied by such animals in all the places service animals are allowed under state law. The law requires that the animal wear a leash, harness, or cape identifying it as a service animal in training.[21]

Responsibility for administering this state law and registering service animals has been delegated to the Client Assistance Program (CAP).[22] As the statute directs, registration of service animals and service animals in training is free.[23] The law directs DHHS to adopt regulations governing the registration of service animals, but as of early 2008 none had been proposed or adopted.[24]

---

disability-related complaints); *see also* http://airconsumer.ost.dot.gov/publications/flyrights.htm#complain (providing an address for submitting complaints) (last visited July 10, 2007).

18. G.S. 168-4.2(a).

19. *Id.*

20. *Id.*

21. G.S. 168-4.2(b).

22. Contact information for the program is available at http://cap.state.nc.us/contactus.htm (last visited July 10, 2007). Funding for the program is managed by the DHHS Division of Vocational Rehabilitation.

23. G.S. 168-4.3.

24. *Id.*

State law also provides several penalties for crimes related to assistance animals. First, it is a Class 3 misdemeanor to deprive a person with a disability of rights granted under the state law, including the right to be accompanied by an assistance animal.[25] It is also a misdemeanor to disguise an animal as a service animal or a service animal in training.[26] In addition, it is a crime for a person to hurt or interfere with an animal knowing that it is an assistance animal; specifically, it is a crime to

- willfully kill the animal (Class H felony);
- willfully cause or attempt to cause serious harm to the animal (Class I felony);
- willfully cause or attempt to cause harm to the animal (Class 1 misdemeanor); or
- willfully taunt, tease, harass, delay, obstruct, or attempt to delay or obstruct the animal in the performance of its assistance duties (Class 2 misdemeanor).[27]

A person convicted under any of these four provisions will not only be sentenced according to the state's structured sentencing guidelines (see Appendix B) but will also be required to make restitution; he or she will have to pay, for example, any veterinary bills, expenses incurred in retraining the animal, and wages lost by the person with a disability while the animal is being retrained.

---

25. G.S. 168-4.5.

26. *Id.*

27. G.S. 14-163.1(a1)–(d). Subsection (a1), which makes it a Class H felony to willfully kill an assistance animal, was added to the statute in 2007. This provision became effective December 1, 2007. S. L. 2007-80 (S 34; amending G.S. 14-163.1 and 15A-1340.16(d)).

Relevant Statutes

State

Article 1 of Chapter 168
Rights.

. . .

§ 168-3. Right to use of public conveyances, accommodations, etc.
Persons with disabilities are entitled to accommodations, advantages, facilities, and privileges of all common carriers, airplanes, motor vehicles, railroad trains, motor buses, streetcars, boats, or any other public conveyances or modes of transportation; hotels, lodging places, places of public accommodation, amusement or resort to which the general public is invited, subject only to the conditions and limitations established by law and applicable alike to all persons.

§ 168-4.2. May be accompanied by service animal.

(a) Every person with a disability has the right to be accompanied by a service animal trained to assist the person with his or her specific disability in any of the places listed in G.S. 168-3, and has the right to keep the service animal on any premises the person leases, rents, or uses. The person qualifies for these rights upon the showing of a tag, issued by the Department of Health and Human Services, under G.S. 168-4.3, stamped "NORTH CAROLINA SERVICE ANIMAL PERMANENT REGISTRATION" and stamped with a registration number, or upon a showing that the animal is being trained or has been trained as a service animal. The service animal may accompany a person in any of the places listed in G.S. 168-3.

(b) An animal in training to become a service animal may be taken into any of the places listed in G.S. 168-3 for the purpose of training when the animal is accompanied by a person who is training the service animal and the animal wears a collar and leash, harness, or cape that identifies the animal as a service animal in training. The trainer shall be liable for any damage caused by the animal while using a public conveyance or on the premises of a public facility or other place listed in G.S. 168-3.

§ 168-4.3. Training and registration of service animal.
The Department of Health and Human Services, shall adopt rules for the registration of service animals and shall issue registrations to a person with a disability who makes application for registration of an animal that serves as a service animal or to a person who is training an animal as a service animal. The rules adopted regarding registration shall require that the animal be trained or

be in training as a service animal. The rules shall provide that the certification and registration need not be renewed while the animal is serving or training with the person applying for the registration. No fee may be charged the person for the application, registration, tag, or replacement in the event the original is lost. The Department of Health and Human Services may, by rule, issue a certification or accept the certification issued by the appropriate training facilities.

§ 168-4.4. Responsibility for service animal.
Neither a person with a disability who is accompanied by a service animal, nor a person who is training a service animal, may be required to pay any extra compensation for the animal. The person has all the responsibilities and liabilities placed on any person by any applicable law when that person owns or uses any animal, including liability for any damage done by the animal.

§ 168-4.5. Penalty.
It is unlawful to disguise an animal as a service animal or service animal in training. It is unlawful to deprive a person with a disability or a person training a service animal of any rights granted the person pursuant to G.S. 168-4.2 through G.S. 168-4.4, or of any rights or privileges granted the general public with respect to being accompanied by animals or to charge any fee for the use of the service animal. Violation of this section shall be a Class 3 misdemeanor.

. . .

---

§ 168A-3. Definitions.

. . .

(7a) "Person with a disability" means any person who (i) has a physical or mental impairment which substantially limits one or more major life activities; (ii) has a record of such an impairment; or (iii) is regarded as having such an impairment. As used in this subdivision, the term:

a. "Physical or mental impairment" means (i) any physiological disorder or abnormal condition, cosmetic disfigurement, or anatomical loss, caused by bodily injury, birth defect or illness, affecting one or more of the following body systems: neurological; musculoskeletal; special sense organs; respiratory, including speech organs; cardiovascular; reproductive; digestive; genitourinary; hemic and lymphatic; skin; and endocrine; or (ii) any mental disorder, such as mental retardation, organic brain syndrome, mental illness, specific learning disabilities, and other developmental disabilities, but (iii) excludes (A) sexual preferences; (B) active alcoholism or drug addiction or abuse; and (C) any disorder, condition or disfigurement which is temporary in nature leaving no residual impairment.

b. "Major life activities" means functions such as caring for one's self, performing manual tasks, walking, seeing, hearing, speaking, breathing, learning, and working.

c. "Has a record of such an impairment" means has a history of, or has been misclassified as having, a mental or physical impairment that substantially limits major life activities.

d. "Is regarded as having an impairment" means (i) has a physical or mental impairment that does not substantially limit major life activities but that is treated as constituting such a limitation; (ii) has a physical or mental impairment that substantially limits major life activities because of the attitudes of others; or (iii) has none of the impairments defined in paragraph a. of this subdivision but is treated as having such an impairment.

. . .

§ 14-163.1. Assaulting a law enforcement agency animal or an assistance animal.

(a) The following definitions apply in this section:

(1) Assistance animal. – An animal that is trained and may be used to assist a "person with a disability" as defined in G.S. 168A-3. The term "assistance animal" is not limited to a dog and includes any animal trained to assist a person with a disability as provided in Article 1 of Chapter 168 of the General Statutes.

(2) Law enforcement agency animal. – An animal that is trained and may be used to assist a law enforcement officer in the performance of the officer's official duties.

(3) Harm. – Any injury, illness, or other physiological impairment; or any behavioral impairment that impedes or interferes with duties performed by a law enforcement agency animal or an assistance animal.

(4) Serious harm. – Harm that does any of the following:

a. Creates a substantial risk of death.

b. Causes maiming or causes substantial loss or impairment of bodily function.

c. Causes acute pain of a duration that results in substantial suffering.
d. Requires retraining of the law enforcement agency animal or assistance animal.
e. Requires retirement of the law enforcement agency animal or assistance animal from performing duties.

(a1) Any person who knows or has reason to know that an animal is a law enforcement agency animal or an assistance animal and who willfully kills the animal is guilty of a Class H felony.

(b) Any person who knows or has reason to know that an animal is a law enforcement agency animal or an assistance animal and who willfully causes or attempts to cause serious harm to the animal is guilty of a Class I felony.

(c) Unless the conduct is covered under some other provision of law providing greater punishment, any person who knows or has reason to know that an animal is a law enforcement agency animal or an assistance animal and who willfully causes or attempts to cause harm to the animal is guilty of a Class 1 misdemeanor.

(d) Unless the conduct is covered under some other provision of law providing greater punishment, any person who knows or has reason to know that an animal is a law enforcement agency animal or an assistance animal and who willfully taunts, teases, harasses, delays, obstructs, or attempts to delay or obstruct the animal in the performance of its duty as a law enforcement agency animal or assistance animal is guilty of a Class 2 misdemeanor.

(d1) A defendant convicted of a violation of this section shall be ordered to make restitution to the person with a disability, or to a person, group, or law enforcement agency who owns or is responsible for the care of the law enforcement agency animal for any of the following as appropriate:

(1) Veterinary, medical care, and boarding expenses for the assistance animal or law enforcement animal.
(2) Medical expenses for the person with the disability relating to the harm inflicted upon the assistance animal.
(3) Replacement and training or retraining expenses for the assistance animal or law enforcement animal.
(4) Expenses incurred to provide temporary mobility services to the person with a disability.
(5) Wages or income lost while the person with a disability is with the assistance animal receiving training or retraining.
(6) The salary of the law enforcement agency animal handler as a result of the lost services to the agency during the time the

handler is with the law enforcement agency animal receiving training or retraining.
(7) Any other expense reasonably incurred as a result of the offense.

(e) This section shall not apply to a licensed veterinarian whose conduct is in accordance with Article 11 of Chapter 90 of the General Statutes.
(f) Self-defense is an affirmative defense to a violation of this section.
(g) Nothing in this section shall affect any civil remedies available for violation of this section.

Federal

28 C.F.R. § 36.104. Definitions.
For purposes of this part, the term—

. . .

Place of public accommodation means a facility, operated by a private entity, whose operations affect commerce and fall within at least one of the following categories—

(1) An inn, hotel, motel, or other place of lodging, except for an establishment located within a building that contains not more than five rooms for rent or hire and that is actually occupied by the proprietor of the establishment as the residence of the proprietor;
(2) A restaurant, bar, or other establishment serving food or drink;
(3) A motion picture house, theater, concert hall, stadium, or other place of exhibition or entertainment;
(4) An auditorium, convention center, lecture hall, or other place of public gathering;
(5) A bakery, grocery store, clothing store, hardware store, shopping center, or other sales or rental establishment;
(6) A laundromat, dry-cleaner, bank, barber shop, beauty shop, travel service, shoe repair service, funeral parlor, gas station, office of an accountant or lawyer, pharmacy, insurance office, professional office of a health care provider, hospital, or other service establishment;
(7) A terminal, depot, or other station used for specified public transportation;
(8) A museum, library, gallery, or other place of public display or collection;
(9) A park, zoo, amusement park, or other place of recreation;

(10) A nursery, elementary, secondary, undergraduate, or postgraduate private school, or other place of education;
(11) A day care center, senior citizen center, homeless shelter, food bank, adoption agency, or other social service center establishment; and
(12) A gymnasium, health spa, bowling alley, golf course, or other place of exercise or recreation.

. . .

Service animal means any guide dog, signal dog, or other animal individually trained to do work or perform tasks for the benefit of an individual with a disability, including, but not limited to, guiding individuals with impaired vision, alerting individuals with impaired hearing to intruders or sounds, providing minimal protection or rescue work, pulling a wheelchair, or fetching dropped items.

28 C.F.R. § 36.302. Modifications in policies, practices, or procedures.

(a) General. A public accommodation shall make reasonable modifications in policies, practices, or procedures, when the modifications are necessary to afford goods, services, facilities, privileges, advantages, or accommodations to individuals with disabilities, unless the public accommodation can demonstrate that making the modifications would fundamentally alter the nature of the goods, services, facilities, privileges, advantages, or accommodations.

(b) Specialties —(1) General. A public accommodation may refer an individual with a disability to another public accommodation, if that individual is seeking, or requires, treatment or services outside of the referring public accommodation's area of specialization, and if, in the normal course of its operations, the referring public accommodation would make a similar referral for an individual without a disability who seeks or requires the same treatment or services.

(2) Illustration—medical specialties. A health care provider may refer an individual with a disability to another provider, if that individual is seeking, or requires, treatment or services outside of the referring provider's area of specialization, and if the referring provider would make a similar referral for an individual without a disability who seeks or requires the same treatment or services. A physician who specializes in treating only a particular condition cannot refuse to treat an individual with a disability for that condition, but is not required to treat the individual for a different condition.

(c) Service animals —(1) General. Generally, a public accommodation shall modify policies, practices, or procedures to permit the use of a service animal by an individual with a disability.

(2) Care or supervision of service animals. Nothing in this part requires a public accommodation to supervise or care for a service animal.

(d) Check-out aisles. A store with check-out aisles shall ensure that an adequate number of accessible check-out aisles are kept open during store hours, or shall otherwise modify its policies and practices, in order to ensure that an equivalent level of convenient service is provided to individuals with disabilities as is provided to others. If only one check-out aisle is accessible, and it is generally used for express service, one way of providing equivalent service is to allow persons with mobility impairments to make all their purchases at that aisle.

Chapter 10

Miscellaneous Issues

- **Pet licensing**
- **Petting zoos**
- **Disposition of dead animals**
- **Emergencies (PETS 2006)**
- **Bird sanctuaries**
- **Pets in hotels**
- **Electronic dog collars**

Pet Licensing

Local governments have the authority to require citizens to obtain licenses and pay taxes on their domestic pets.[1] Several jurisdictions in North Carolina have adopted ordinances imposing such licensing requirements. Their form and purpose may vary according to the city's or county's desire to promote other policy goals, such as encouraging owners to sterilize or vaccinate their animals or expanding the authority to impound stray animals.

In Durham County, for example, pet owners pay a $10 license fee for each sterile animal and a $75 fee for an unsterilized animal.[2] Asheville takes a slightly different approach, charging a flat licensing fee of $10 for all animals but requiring animal owners who have not had their pets sterilized to apply for an "unaltered animal permit" and pay a one-time $100 fee.[3] A recent study conducted by North Carolina State University veterinary graduate students examined Asheville's program and concluded that it did increase the number of sterilized pets in the city.[4] Another study by a University of North Carolina

---

1. N.C. Gen. Stat. 153A-153 (counties) (hereinafter G.S.); G.S. 160A-212 (cities).

2. See Durham County Animal Control Division, Vaccination and Licensing, www.durhamcountync.gov/departments/anml/Vaccination_and_Licensing.html (last visited Oct. 12, 2007).

3. Asheville Code of Ordinances, § 3-5, www.buncombecounty.org/governing/depts/Sheriff/animalControl.asp

4. Courtney Pierce and Jennifer Reed, "Preliminary Assessment of the Effects of Recent Spay/Neuter Legislation in Buncombe County, NC" (June 2007) (graduate student research paper), www.ncanimalcontrol.unc.edu/pdfs/FinalReportPierceReed.pdf (last visited Oct. 12, 2007).

graduate student in public administration looked at jurisdictions that linked licensing with rabies vaccination requirements. The author concluded that linking the two requirements not only increased the local vaccination rates and improved enforcement of licensure requirements but could also significantly increase the number of animals licensed.[5]

Some local governments have managed to generate significant income streams from license taxes. For example, New Hanover County, which started its mandatory licensing program in 1999, has been able to gradually increase its revenues every year. In fiscal year 2006–2007, the program generated over $725,000.[6] Other jurisdictions, however, have concluded that the political ramifications of imposing a new tax, the likelihood that only the "good, responsible citizens would pay," and the administrative burden related to collection outweigh the potential benefits of such a tax.[7]

Petting Zoos

In the fall of 2004, over a hundred people—mostly children under the age of six—contracted a communicable E. coli infection after visiting the petting zoo at the North Carolina State Fair.[8] At the time, North Carolina had no state laws regulating sanitation at petting zoos. In response to the E. coli outbreak, the North Carolina Department of Agriculture and Consumer Services (the Department)—which oversees the state fair—instituted new restrictions at the fair's animal exhibitions.[9] In addition, the General Assembly passed legislation directing the commissioner of agriculture to establish a permitting system for animal exhibitions.

The new law defines *animal exhibitions* as agricultural fairs where animals are displayed on the exhibition grounds, where they may have physical contact with humans. The term *fair* is a specialized term that refers to exhibitions designed to promote agriculture and other industries by offering premiums and

---

5. Catherine M. Clark, "The Truth about Cats and Dogs: Vaccinations, Licenses, Service, Revenue," *Popular Government* 67 (Winter 2002): 40.

6. Conversation with Jean McNeil, director of Animal Control for New Hanover County (July 16, 2007).

7. *See, e.g.*, Nate DeGraff, "Paying to Own a Pet," *Greensboro News-Record*, Sept. 24, 2006.

8. *See* Centers for Disease Control and Prevention (CDC), "Outbreaks of Escherichia coli 0157:H7 Associated with Petting Zoos—North Carolina, Florida, and Arizona, 2004 and 2005," *Morbidity and Mortality Weekly* (*MMWR*), 54(50): 1277–1280 (Dec. 23, 2005); Lisa Hoppenjans, "As Girl Copes, Legacy May Protect Others," *Raleigh News & Observer*, July 26, 2005, 1A.

9. *See MMWR*, 54(50): 1280.

awards.[10] Such fairs are already required to obtain licenses from the Department; the sanitation requirements related to animal exhibitions add a new layer of regulation.[11] The regulations, which went into effect in September 2006, require exhibitions to

- provide fencing to minimize contact between the public and the manure or bedding of the animals;
- provide hand-washing stations (which should include soap and running water, not hand sanitizers and hand wipes);
- post signs regarding the health risks related to animal contact and identifying the location of hand-washing stations; and
- maintain health certificates for animals included in the exhibition.[12]

In July 2007 the National Association of State Public Health Veterinarians (NASPHV) released a report recommending measures for preventing disease associated with animals in public settings.[13] The report's recommendations are largely consistent with the new state regulations.

If an exhibition is in violation of the regulations, the Department may deny, suspend, or revoke its permit and may also assess a civil monetary penalty of up to five thousand dollars.[14] In addition, private individuals who are harmed at such exhibitions may consider bringing civil lawsuits against the exhibition operators to recover money damages. In 2007, however, legislation that restricts the ability of private individuals to recover damages was enacted.[15] The new law provides that, subject to limited exceptions, the operator of an exhibition will not be liable for "injury to or death of a participant resulting from the inherent risks" related to the exhibition's activity.[16] To take advantage of this limitation on liability, the operator must post a sign warning the public about the "inherent risks" related to the animal exhibition.[17]

---

10. G.S. 106-520.1. The term *fair* does not encompass "noncommercial community fairs." G.S. 106-520.3.

11. G.S. 106-520.3A.

12. N.C. Admin. Code tit. 2, ch. 52K, §§ .0101–.0702 (hereinafter N.C.A.C.

13. *See* CDC, "Compendium of Measures to Prevent Disease Associated with Animals in Public Settings," 2007, *MMWR* 56(RR-5): 1–28 (July 6, 2007), www.cdc.gov/mmwr/PDF/rr/rr5605.pdf (last visited Oct. 12, 2007).

14. G.S. 106-520.3(f).

15. S.L. 2007-171 (amending G.S. 99E-30(1)).

16. G.S. 99E-31(a). The exceptions may apply if the operator (1) commits an act or omission that constitutes negligence or willful or wanton disregard for the safety of the participant, or (2) has actual knowledge or reasonably should have known of a dangerous condition. G.S. 99E-31(b).

17. G.S. 99E-32. The sign must include the following language: "WARNING. Under North Carolina law, there is no liability for an injury to or death of a participant in an agritourism activity conducted at this agritourism location if such injury or death

Disposal of Dead Animals

With respect to dead animals, state law imposes various duties on animal owners, owners of property where dead animals are found, and cities and counties, as well as on the North Carolina Department of Transportation (NCDOT). If a domesticated animal dies, the animal's owner or the person who owns or operates the property where the animal died must bury it within twenty-four hours of learning of its death. The animal must be buried at least three feet deep and no less than 300 feet from any flowing stream or public body of water. Alternatively, the responsible person may contact the state veterinarian at the Department of Agriculture to seek approval for another method of disposal.[18] A willful violation of this law is a Class 2 misdemeanor.[19] Cities and counties are required to designate a person to arrange for the removal of dead animals whose owners cannot be identified. Cities are responsible for animals within the city limits, and counties are responsible for all areas outside the limits of any municipality.[20] Some jurisdictions, such as Winston-Salem and Asheville, offer to pick up dead animals and dispose of them.[21]

NCDOT is charged with removing and disposing of dead animals from primary and secondary roads. If it finds some evidence about the ownership of a dog found dead, NCDOT must take "reasonable steps" to notify the owner.[22] Some local governments also play a role in removing dead animals from the roads within their jurisdictions.[23]

---

results from the inherent risks of the agritourism activity. Inherent risks of agritourism activities include, among others, risks of injury inherent to land, equipment, and animals, as well as the potential for you to act in a negligent manner that may contribute to your injury or death. You are assuming the risk of participating in this agritourism activity." *Id. See also* G.S. 520 and G.S. 520.3A.

18. G.S. 106-403.

19. G.S. 106-405.

20. G.S. 106-403.

21. *See* City of Winston-Salem, Sanitation Division, www.cityofws.org/Home/Departments/Sanitation/Collections ("The City provides dead animal collection Monday through Friday from 8:00 am to 2:00 pm and on Saturdays from 8:00 am to 12:00 pm. Animals are collected from the streets. No collections are made on private property. Residents should place animals in a bag and place by the curb.") (last visited July 13, 2007); City of Asheville, Public Works Department, www.ashevillenc.gov/residents/public_services/sanitation/default.aspx?id=750 ("Small dead animals, such as a cat, dog or other small, household pet, must be wrapped in a plastic bag and placed at the curb.") (last visited July 13, 2007).

22. G.S. 136-81(21).

23. For example, the City of Durham will collect and dispose of dead animals found on streets within the city limits. City of Durham INFO-Line, Dead Animal Removal (contacted July 12, 2007).

Emergency Preparedness

In 2006 Congress passed federal legislation addressing care for animals during natural disasters and emergencies.[24] The federal law makes three policy changes. First, it requires state and local government emergency preparedness plans to "take into account the needs of individuals with household pets and service animals prior to, during, and following a major disaster or emergency."[25] These plans are important because the Federal Emergency Management Agency (FEMA) may rely on them when allocating and distributing certain federal preparedness funds to the states and other governmental entities.[26] The second change specifically allows FEMA to provide funding to state and local authorities for animal emergency preparedness purposes.[27] Finally, the act authorizes FEMA to provide assistance and funding to state and local governments involved with the "provision of rescue, care, shelter, and essential needs" of people who have household pets and service animals.[28] In October 2007 FEMA released guidance outlining the parameters of the new policy.[29] In summary, the guidance

- defines key terms, including *household pet* and *service animal*;
- identifies entities that are eligible for reimbursement;
- lists the types of activities and services that are reimbursable, including certain costs incurred for labor, facilities, supplies, equipment, veterinary services, transportation, and removal and disposal of dead animals.

Bird Sanctuaries

In North Carolina, only cities have the authority to adopt ordinances establishing bird sanctuaries within their jurisdictions.[30] If a city establishes a sanctuary, it may restrict the hunting and trapping of birds within the city limits, but it may not protect birds that are considered pests under state law. For example, the North Carolina Pesticide Board has declared the red-winged blackbird a pest and has authorized the use of pesticides on the birds in certain

---

24. Pets Evacuation and Transportation Standards Act of 2006 (PETS Act), P.L. No. 109-308, 120 Stat. 1725 (2006).
25. 42 U.S.C. § 5196b(g).
26. 42 U.S.C. § 5196b(a), (f).
27. 42 U.S.C. § 5196(2).
28. 42 U.S.C. § 5170b(a)(3).
29. FEMA Disaster Assistance Policy 9523.19, Eligible Costs Related to Pet Evacuation and Sheltering (Oct. 24, 2007), www.fema.gov/government/grant/pa/9523.19.shtm.
30. G.S. 160A-188.

circumstances.[31] Use of such a pesticide would not be prohibited within a bird sanctuary. Below is an example of a sanctuary ordinance:

> (a) *Town designated as sanctuary.* The area within the corporate limits of the town and all land owned or leased by the town outside the corporate limits is hereby designated as a bird sanctuary, as authorized by G.S. 160A-188.
>
> (b) *Unlawful to trap, etc.* It shall be unlawful intentionally to trap, hunt, shoot, or otherwise kill, within the sanctuary hereby established, any native wild bird, except those birds classified as a pest under article 22A of chapter 113 of the General Statutes (G.S. 113-300.1 et seq.) and the Structural Pest Control Act of North Carolina of 1955 (G.S. 106-55.22 et seq.) or the North Carolina Pesticide Law of 1971 (G.S. 143-434 et seq.), pursuant to an appropriate permit issued by the North Carolina Wildlife Commission.[32]

Pets in Hotels

Operators of inns and hotels may establish policies allowing guests to bring pets into sleeping rooms and adjoining rooms.[33] Hotels that do allow pets must (1) post a notice of that fact in the registration area, (2) post a sign in any sleeping room where pets are allowed, and (3) prohibit pets from at least 10 percent of the sleeping rooms. An operator may be charged with a Class 3 misdemeanor for failing to comply with these requirements. In addition, a person who brings a pet into a room where pets are prohibited may be charged with a Class 3 misdemeanor. This law, however, does *not* apply to the admittance of assistance animals to inns or hotels.

Electronic Dog Collars

It is a crime to intentionally remove or destroy an electronic dog collar or other electronic device placed on a dog by its owner to maintain control of the dog. The first violation is a Class 3 misdemeanor, and subsequent convictions are Class 2 misdemeanors.

---

31. 2 N.C.A.C. 09L .0706 ("Pesticides registered for use to control the red-winged blackbird may be used when it is committing or about to commit depredations upon ornamental or shade trees, agricultural crops, livestock, or wildlife, or when concentrated in such numbers or manner as to constitute a health hazard or other nuisance.").

32. Town of Smithsfield Ordinance No. 427, 2-7-06.

33. G.S. 72-7.1.

Relevant Statutes

§ 14-401.17. Unlawful removal or destruction of electronic dog collars.

(a) It is unlawful to intentionally remove or destroy an electronic collar or other electronic device placed on a dog by its owner to maintain control of the dog.

(b) A first conviction for a violation of this section is a Class 3 misdemeanor. A second or subsequent conviction for a violation of this section is a Class 2 misdemeanor.

(c) This act is enforceable by officers of the Wildlife Resources Commission, by sheriffs and deputy sheriffs, and peace officers with general subject matter jurisdiction.

(d) Repealed by Session Laws 2005-94, s. 1, effective December 1, 2005, and applicable to offenses committed on or after that date.

§ 72-7.1. Admittance of pets to hotel rooms.

(a) Innkeepers may permit pets in rooms used for sleeping purposes and in adjoining rooms. Persons bringing pets into a room in which they are not permitted are in violation of this section and punishable according to subsection (d) of this section.

(b) Innkeepers allowing pets must post a sign measuring not less than five inches by seven inches at the place where guests register informing them pets are permitted in sleeping rooms and in adjoining rooms. If certain pets are permitted or prohibited, the sign must so state. If any pets are permitted, the innkeeper must maintain a minimum of ten percent (10%) of the sleeping rooms in the inn or hotel as rooms where pets are not permitted and the sign required by this subsection must also state that such rooms are available.

(c) All sleeping rooms in which the innkeeper permits pets must contain a sign measuring not less than five inches by seven inches, posted in a prominent place in the room, which shall be separate from the sign required by G.S. 72-6, stating that pets are permitted in the room, or whether certain pets are prohibited or permitted in the room, and stating that bringing pets into a room in which they are not permitted is a Class 3 misdemeanor.

(d) Any person violating the provisions of this section shall be guilty of a Class 3 misdemeanor.

(e) The provisions of this section are not applicable to assistance dogs admitted to sleeping rooms and adjoining rooms under the provisions of Chapter 168 of the General Statutes.

Article 4 of Chapter 99E

§ 99E-30. Definitions.

As used in this Article, the following terms mean:

(1) Agritourism activity. – Any activity carried out on a farm or ranch that allows members of the general public, for recreational, entertainment, or educational purposes, to view or enjoy rural activities, including farming, ranching, historic, cultural, harvest-your-own activities, or natural activities and attractions. An activity is an agritourism activity whether or not the participant paid to participate in the activity. "Agritourism activity" includes an activity involving any animal exhibition at an agricultural fair licensed by the Commissioner of Agriculture pursuant to G.S. 106-520.3.

(2) Agritourism professional. – Any person who is engaged in the business of providing one or more agritourism activities, whether or not for compensation.

(3) Inherent risks of agritourism activity. – Those dangers or conditions that are an integral part of an agritourism activity including certain hazards, including surface and subsurface conditions, natural conditions of land, vegetation, and waters, the behavior of wild or domestic animals, and ordinary dangers of structures or equipment ordinarily used in farming and ranching operations. Inherent risks of agritourism activity also include the potential of a participant to act in a negligent manner that may contribute to injury to the participant or others, including failing to follow instructions given by the agritourism professional or failing to exercise reasonable caution while engaging in the agritourism activity.

(4) Participant. – Any person, other than the agritourism professional, who engages in an agritourism activity.

(5) Person. – An individual, fiduciary, firm, association, partnership, limited liability company, corporation, unit of government, or any other group acting as a unit.

§ 99E-31. Liability.

(a) Except as provided in subsection (b) of this section, an agritourism professional is not liable for injury to or death of a participant resulting from the inherent risks of agritourism activities, so long as the warning contained in G.S. 99E-32 is posted as required and, except as provided in subsection (b) of this section, no participant or participant's representative can maintain an action against or recover from an agritourism professional for injury, loss, damage, or death of the participant resulting exclusively from any of the inherent risks of agritourism activities. In any action for damages against an agritourism professional for agritourism activity, the agritourism professional must plead the affirmative defense of assumption of the risk of agritourism activity by the participant.

(b) Nothing in subsection (a) of this section prevents or limits the liability of an agritourism professional if the agritourism professional does any one or more of the following:

(1) Commits an act or omission that constitutes negligence or willful or wanton disregard for the safety of the participant, and that act or omission proximately causes injury, damage, or death to the participant.

(2) Has actual knowledge or reasonably should have known of a dangerous condition on the land, facilities, or equipment used in the activity or the dangerous propensity of a particular animal used in such activity and does not make the danger known to the participant, and the danger proximately causes injury, damage, or death to the participant.

(c) Nothing in subsection (a) of this section prevents or limits the liability of an agritourism professional under liability provisions as set forth in Chapter 99B of the General Statutes.

(d) Any limitation on legal liability afforded by this section to an agritourism professional is in addition to any other limitations of legal liability otherwise provided by law.

§ 99E-32. Warning required.

(a) Every agritourism professional must post and maintain signs that contain the warning notice specified in subsection (b) of this section. The sign must be placed in a clearly visible location at the entrance to the agritourism location and at the site of the agritourism activity. The warning notice must consist of a sign in black letters, with each letter to be a minimum of one inch in height. Every written contract entered into by an agritourism professional for the providing of professional services, instruction, or the rental of equipment to a participant, whether or not the contract involves agritourism activities on or off the location or at the site of the agritourism activity, must contain in clearly readable print the warning notice specified in subsection (b) of this section.

(b) The signs and contracts described in subsection (a) of this section must contain the following notice of warning:

"WARNING
Under North Carolina law, there is no liability for an injury to or death of a participant in an agritourism activity conducted at this agritourism location if such injury or death results from the inherent risks of the agritourism activity. Inherent risks of agritourism activities include, among others, risks of injury inherent to land, equipment, and animals, as well as the potential for you to act in a negligent manner that may contribute to your injury or death. You are assuming the risk of participating in this agritourism activity."

(c) Failure to comply with the requirements concerning warning signs and notices provided in this subsection will prevent an agritourism professional from invoking the privileges of immunity provided by this Article.

---

§ 106-403. (Effective until October 1, 2009) Disposition of dead domesticated animals.

It is the duty of the owner of domesticated animals that die from any cause and the owner or operator of the premises upon which any domesticated animals die, to bury the animals to a depth of at least three feet beneath the surface of the ground within 24 hours after knowledge of the death of the domesticated animals, or to otherwise dispose of the domesticated animals in a manner approved by the State Veterinarian. It is a violation of this section to bury any dead domesticated animal closer than 300 feet to any flowing stream or public body of water. It is unlawful for any person to remove the carcasses of dead domesticated animals from the person's premises to the premises of any other person without the written permission of the person having charge of the other premises and without burying the carcasses as provided under this section. The governing body of each municipality shall designate some appropriate person whose duty it shall be to provide for the removal and disposal, according to the provisions of this section, of any dead domesticated animals located within the limits of the municipality when the owner of the animals cannot be determined. The board of commissioners of each county shall designate some appropriate person whose duty it shall be to provide for the removal and disposal under this section, of any dead domesticated animals located within the limits of the county, but without the limits of any municipality, when the owner of the animals cannot be determined. All costs incurred by a municipality or county in the removal of dead domesticated animals shall be recoverable from the owner of the animals upon admission of ownership or conviction. "Domesticated animal" as used in this section includes poultry.

§ 106-403. (Effective October 1, 2009) Disposition of dead domesticated animals.

It shall be the duty of the owner or person in charge of any of his domesticated animals that die from any cause and the owner, lessee, or person in charge of any land upon which any domesticated animals die, to bury the same to a depth of at least three feet beneath the surface of the ground within 24 hours after knowledge of the death of said domesticated animals, or to otherwise dispose of the same in a manner approved by the State Veterinarian. It shall be a violation of this statute to bury any dead domesticated animal closer than 300 feet to any flowing stream or public body of water. It shall be unlawful for any person to remove the carcasses of dead domesticated animals from his premises to the premises of any other person without the written permission of the person having charge of such premises and without burying said carcasses as above provided. The governing body of each municipality shall designate some appropriate person whose duty it shall be to provide for the removal and disposal, according to the provisions of this section, of any dead domesticated animals located within the limits of the municipality when the owner or owners of said animals cannot be determined. The board of commissioners of each county shall designate some appropriate person whose duty it shall be to provide for the removal and disposal, according to the provisions of this section, of any dead domesticated animals located within the limits of the county, but without the limits of any municipality, when the owner or owners of said animals cannot be determined. All costs incurred by a municipality or county in the removal of a dead domesticated animal shall be recoverable from the owner of such animal upon admission of ownership or conviction. "Domesticated animal" as used herein shall include poultry.

§ 106-520.3A. Animal exhibition regulation; permit required; civil penalties.

(a) Title. – This section may be referred to as "Aedin's Law". This section provides for the regulation of animal exhibitions as they may affect the public health and safety.

(b) Definitions. – As used in this section, unless the context clearly requires otherwise:

(1) "Animal" means only those animals that may transmit infectious diseases.

(2) "Animal exhibition" means any sanctioned agricultural fair where animals are displayed on the exhibition grounds for physical contact with humans.

(c) Permit Required. – No animal exhibition may be operated for use by the general public unless the owner or operator has obtained an operation permit issued by the Commissioner. The Commissioner may issue an operation permit only after physical inspection of the animal exhibition and a determination that the animal exhibition meets the requirements of this section and rules adopted pursuant to this section. The Commissioner may deny, suspend, or revoke a permit on the basis that the exhibition does not comply with this section or rules adopted pursuant to this section.

(d) Rules. – For the protection of the public health and safety, the Commissioner of Agriculture, with the advice and approval of the State Board of Agriculture, and in consultation with the Division of Public Health of the Department of Health and Human Services, shall adopt rules concerning the operation of and issuance of permits for animal exhibitions. The rules shall include requirements for:

(1) Education and signage to inform the public of health and safety issues.
(2) Animal areas.
(3) Animal care and management.
(4) Transition and nonanimal areas.
(5) Hand-washing facilities.
(6) Other requirements necessary for the protection of the public health and safety.

(e) Educational Outreach. – The Department shall continue its consultative and educational efforts to inform agricultural fair operators, exhibitors, agritourism business operators, and the general public about the health risks associated with diseases transmitted by physical contact with animals.

(f) Civil Penalty. – In addition to the denial, suspension, or revocation of an operation permit, the Commissioner may assess a civil penalty of not more than five thousand dollars ($5,000) against any person who violates a provision of this section or a rule adopted pursuant to this section. In determining the amount of the penalty, the Commissioner shall consider the degree and extent of harm caused by the violation.

The clear proceeds of civil penalties assessed pursuant to this section shall be remitted to the Civil Penalty and Forfeiture Fund in accordance with G.S. 115C-457.2.

(g) Legal Representation by Attorney General. – It shall be the duty of the Attorney General to represent the Department of Agriculture and Consumer Services or designate a member of the Attorney General's staff to represent the Department in all actions or proceedings in connection with this section.

§ 136-18. Powers of Department of Transportation.

. . .

(21) The Department of Transportation is hereby authorized and directed to remove all dead animals from the traveled portion and rights-of-way of all primary and secondary roads and to dispose of such animals by burial or otherwise. In cases where there is evidence of ownership upon the body of any dead dog, the Department of Transportation shall take reasonable steps to notify the owner thereof by mail or other means.

§ 153A-153. Animal tax.
A county may levy an annual license tax on the privilege of keeping dogs and other pets within the county.

§ 160A-188. Bird sanctuaries.
A city may by ordinance create and establish a bird sanctuary within the city limits. The ordinance may not protect any birds classed as a pest under Article 22A of Chapter 113 of the General Statutes and the Structural Pest Control Act of North Carolina of 1955 or the North Carolina Pesticide Law of 1971. When a bird sanctuary has been established, it shall be unlawful for any person to hunt, kill, trap, or otherwise take any protected birds within the city limits except pursuant to a permit issued by the North Carolina Wildlife Resources Commission under G.S. 113-274(c) (1a) or under any other license or permit of the Wildlife Resources Commission specifically made valid for use in taking birds within city limits.

§ 160A-212. Animal taxes.
A city shall have power to levy an annual license tax on the privilege of keeping any domestic animal, including dogs and cats, within the city. This section shall not limit the city's authority to enact ordinances under G.S. 160A-186.

Appendixes

Appendix A

Euthanasia Methods Approved by the AVMA, AHA, and HSUS

	AHA	HSUS	AVMA
Injection of sodium pentobarbital AHA considers it to be only acceptable method HSUS identifies it as the preferred method HSUS and AVMA impose some conditions on the administration of the drug (i.e., whether it is administered intravenously or through other methods) AVMA indicates that the injection of other barbiturates might also be acceptable (secobarbital)	✓	✓	✓
Injection of sodium pentobarbital combined with another drug HSUS and AVMA approve in some circumstances but both disapprove of combinations of pentobarbital with a neuromuscular blocking agent		✓	✓
Carbon Monoxide Gas AVMA approves only if (a) compressed CO in cylinders is used; (b) certain precautions are taken (such as appropriate training for personnel and the use of a chamber that allows for individual separation of animals); and (c) not used in animals less than 16 weeks old (except to induce loss of consciousness). HSUS considers the use of CO to be conditionally acceptable in those states (like North Carolina) where shelters do not have direct access to sodium pentobarbital. HSUS does not approve of its use for animals who are geriatric, under four months of age, sick, injured or obviously pregnant		✓	✓
Other inhalant agents AVMA approves of the use of some other inhalant agents (such as halothane, enflurane, isoflurane, sevoflurane, mehoxyflurane, and desflurane) for euthanasia of small animals (<7 kg). AVMA approves of the use of compressed carbon dioxide in cylinders but only if certain conditions are satisfied. AVMA conditionally approves of the use of nitrogen and argon gases in certain situations, but states that other methods of euthanasia are preferable. AVMA does not recommend using any of these inhalant agents alone in animals less than 16 weeks of age.			✓
Gunshot HSUS approves of gunshot as a method of euthanasia only in an emergency field situation where (a) an animal cannot be confined and transferred to the shelter, (b) sodium pentobarbital is unavailable, and (c) the personnel are appropriately trained. AVMA conditionally approves of using a gunshot but indicates that the method should not be used for routine euthanasia of animals in animal control situations (such as shelters).		✓	✓
Injection of potassium chloride AVMA approves of the injection of potassium chloride but only when the animal is under general anesthesia			✓
Penetrating captive bolt AVMA approves of the use of a penetrating captive bolt to the head for euthanasia of dogs in limited situations (e.g., research facilities, farms when the use of drugs is inappropriate)			✓
Electrocution AVMA approves of the use of electrocution for euthanasia of dogs but only in very limited circumstances.			✓

AHA = American Humane Association HSUS = Humane Society of the United States AVMA = American Veterinary Medicine Assocation

Appendix B

Punishment under Structured Sentencing

Generally

This appendix briefly describes the application of structured sentencing to both felonies and misdemeanors so that interested readers can learn what a particular felony or misdemeanor classification means in terms of actual punishments for individual defendants. Readers interested in a more complete discussion should consult Stevens H. Clarke, *Law of Sentencing, Probation, and Parole in North Carolina* (UNC Institute of Government, 2d ed. 1997), and the supplement to that book, *Administration of Justice Bulletin* No. 99/01 (UNC Institute of Government, 1999); John Rubin, Ben F. Loeb Jr. and James C. Drennan, *Punishments for North Carolina Crimes and Motor Vehicle Offenses* (UNC School of Government, 2005, and 2006 supplement), from which much of this chapter is drawn; and North Carolina Sentencing and Policy Advisory Commission, *Structured Sentencing Training and Reference Manual* (Revised December 1, 2004). All of these publications are available from the School of Government's Publications Sales Office: telephone 919.966.4119;

This appendix is reproduced with the permission of the author from Jessica Smith, *North Carolina Crimes: A Guidebook on the Element of Crime*, 6th ed. (Chapel Hill: UNC School of Government, 2007).

fax 919.962.2707; e-mail sales@sog.unc.edu; or visit the School's website at www.sog.unc.edu.

Structured sentencing became effective for offenses committed on or after October 1, 1994. The only misdemeanors currently not subject to structured sentencing are impaired driving under G.S. 20-138.1; commercial impaired driving under G.S. 20-138.2; a second or subsequent conviction of a zero tolerance offense under G.S. 20-138.2A (commercial drivers) or G.S. 20-138.2B (school bus and child care vehicle drivers); and failing to comply with health control measures under G.S. 130A-25. The punishment scheme for the motor vehicle offenses just listed is provided in G.S. 20-179 and is discussed in more detail in the note on "Punishment" under "Impared Driving and Related Offenses" in Chapter 28, "Motor Vehicle Offenses." Also exempt from structured sentencing are felony convictions in which defendants are sentenced as violent habitual felons under G.S. 14-7.12.

The United States Supreme Court's 2004 decision in *Blakely v. Washington* (542 U.S. 296) dramatically affected North Carolina's structured sentencing scheme for felonies, as well as its sentencing scheme for impaired driving–related convictions punished under G.S. 20-179. Since then, the North Carolina General Assembly has enacted legislation designed to cure the constitutional defects with the state's felony structured sentencing scheme (S.L. 2005-145). That legislation, however, applies only to felonies committed on or after June 30, 2005 (S.L. 2005-145 sec. 5). Sentences for felonies committed before that date are not subject to the new law's provisions, but they must comply with the *Blakely* decision. S.L. 2006-253 amended G.S. 20-179 to conform to *Blakely.* For a detailed discussion of *Blakely* and its impact on North Carolina sentencing schemes, see Jessica Smith, *North Carolina Sentencing After* Blakely v. Washington *and the* Blakely *Bill* (September 2005), available at www.sog.unc.edu/programs/crimlaw/Blakely%20Update.pdf. Readers need to be aware, however, that the law in this area is still evolving and are advised to consult more recent case law if faced with a *Blakely* question.

Felonies

The centerpiece of structured sentencing for felonies is the statutory table of punishments, commonly referred to as the sentencing grid, in G.S. 15A-1340.17(c) and included as Table 1 at the end of this chapter. Two other tables necessary to determine the appropriate sentence for a felony are also included: Table 2, dealing with maximum sentences, and Table 3, dealing with prior record level.

A sentencing court essentially must take seven steps to determine the appropriate sentence for felonies under structured sentencing. These steps (with the appropriate source to consult) are as follows:

(1) Determine the offense class for each felony conviction (listed under each offense in this book as "Punishment")
(2) Determine the prior record level for the defendant (Table 3)
(3) Consider aggravating and mitigating factors [G.S. 15A-1340.16(d), (e)]
(4) Select a minimum sentence from the applicable minimum sentence range (Table 1)
(5) Determine the maximum sentence (Table 2)
(6) Determine the sentence disposition (active, intermediate, or community) (Table 1)
(7) Consider the appropriateness of restitution (see the discussion later in this chapter)

Each of these steps is summarized below. The main exceptions also are noted below, under the heading "Special Provisions."

The sentencing court will inform the defendant of (1) the active jail or prison term imposed, (2) the active term's minimum and maximum length, and (3) whether that term will be activated or suspended. If the active term is suspended, the defendant will be informed of the type of alternative disposition (intermediate or community) he or she will receive. This procedure will be followed for each offense conviction, unless a particular conviction is consolidated or set to run concurrently with other convictions.

Offense Class

The first step in determining the appropriate sentence for a felony is to identify the class of the felony. There are 10 classes of felonies under structured sentencing: A, B1, B2, C, D, E, F, G, H, and I. Punishments increase in severity as you move from Class I to Class A. For each felony listed in this book, the class of the offense is indicated under the "Punishment" section of the discussion. For example, a violation of G.S. 14-32(c) ("Assault with a Deadly Weapon with Intent to Kill"; see Chapter 7) is a Class E felony.

Prior Record Level

The second step is to determine the defendant's prior record level. Under G.S. 15A-1340.14(c), a defendant is assigned to one of six prior record levels (I through VI) based on the number of points he or she receives under structured sentencing. For example, a defendant with five to eight points is in prior record level III. Punishments increase in severity as you move from prior record level I to VI. The six prior record levels, and the point ranges for each level, are indicated in the prior record level worksheet issued by the Administrative Office of the Courts (AOC-CR-600, October 2006), which appears as Table 3 at the end of this chapter. The felony portion of this form is derived from G.S. 15A-1340.14, while the misdemeanor portion comes from G.S. 15A-1340.21. The form will be used by the prosecutor and the judge to make a record of a defendant's criminal history. Readers of this book should check for updated versions of this form from the AOC before using the example in this book in court. AOC forms are available online at www.nccourts.org/Forms/FormSearch.asp.

Under G.S. 15A-1340.14(b), a defendant is assigned prior record level points in one of three ways. First, if a defendant has any prior convictions, points are assigned to each conviction based on the offense class of the conviction. For example, a prior conviction for a Class H felony is assigned two points. The point values for each offense class are indicated in Table 3. In calculating points for prior convictions, the following structured sentencing rules must be kept in mind.

Prior misdemeanors. In determining a defendant's prior record level for the purpose of felony sentencing, prior convictions of Class 2 and 3 misdemeanors do not count; nor do prior misdemeanor offenses of any class under Chapter 20 of the General Statutes, except misdemeanor death by vehicle under G.S. 20-141.4(a2), impaired driving under G.S. 20-138.1, and commercial impaired driving under G.S. 20-138.2 [G.S. 15A-1340.14(b)].

Current classification of prior conviction. Under the structured sentencing rules for felonies, the classification of the prior offense is the classification assigned to that offense at the time the current offense was committed. For example, if the defendant has a prior conviction for second-degree burglary, the points assigned to the conviction would depend on the current classification of the offense (now Class G), not the classification in effect at the time the conviction occurred [G.S. 15A-1340.14(c)].

Multiple prior convictions. If the defendant was convicted of more than one offense in a single superior court during one calendar week, only the conviction with the highest point total is counted. If the defendant was convicted of more than one offense during a single session (that is, day) of district court, only the most serious conviction is counted [G.S. 15A-1340.14(d)].

Convictions from other jurisdictions. Unless the prosecution or defendant proves otherwise, a conviction from another jurisdiction is classified as a Class I felony if that jurisdiction classifies the offense as a felony. Similarly, unless the prosecution proves otherwise, a conviction from another jurisdiction is classified as a Class 3 misdemeanor if the other jurisdiction classifies the offense as a misdemeanor [G.S. 15A-1340.14(e)].

The second way a defendant is assigned prior record level points is if all the elements of the present offense are included in a prior offense. In essence, the current offense must be the same as or a lesser-included offense of a prior offense committed by the defendant. This fact adds one point and the additional point may be counted whether or not the prior offense was used in determining the prior record level [G.S. 15A-1340.14(b)(6)].

Third, one point is added if the defendant was on probation, parole, or post-release supervision, was serving an active sentence, or was an escapee when the current offense was committed [G.S. 15A-1340.14(b)(7)].

Most prior record level points are based on prior convictions and thus are excepted from the *Blakely* decision, discussed above [G.S. 15A-1340.14(b)(1)–(5)]. As noted above, under G.S. 15A-1340.14(b)(6), one point is assigned when all the elements of the present offense are included in any prior offense for which the defendant was convicted. Although S.L. 2005-145 did not address this point, at least one North Carolina case has held that *Blakely* has no implications for this prior record level point (172 N.C. App. 839). Also as noted above, under G.S. 15A-1340.14(b)(7), one point is assigned if the defendant committed the offense while on probation, parole, or post-release supervision, while serving a sentence of imprisonment, or while an escapee from a correctional institution. Both S.L. 2005-145 and the North Carolina case law treat this point as subject to *Blakely*—meaning that unless admitted to by the defendant, it must be submitted to the jury and proved beyond a reasonable doubt (S.L. 2005-145, sec. 1–2; 172 N.C. App. 839). The final *Blakely* issue with regard to prior record level points pertains to the classification of prior convictions from other jurisdictions under G.S. 15A-1340.14(e). It has been held that the determination of whether a conviction from another jurisdiction is substantially similar to a North Carolina offense is made by a judge, at least when the court does not look beyond statutory definitions to make the determination (175 N.C. App. 250; 175 N.C. App. 492). For more detail on these issues, see Jessica Smith, *North Carolina Sentencing After* Blakely v. Washington *and the* Blakely *Bill* (September 2005), available at www.sog.unc.edu/programs/crimlaw/Blakely%20Update.pdf.

Aggravating and Mitigating Factors

Table 1, the sentencing grid, contains three ranges of punishment for all but one of the 10 felony classes. The exception is Class A, for which punishment is death or life without parole. The three ranges of punishment for all of the other felony classes are mitigated, presumptive, and aggravated.

Mitigated terms are at the low end, aggravated terms are at the high end, and presumptive sentences fall in the middle [G.S. 15A-1340.17(c)(2)–(4)]. The presumptive range is the basic sentencing range—no special findings need to be made for presumptive range sentencing to apply [G.S. 15A-1340.16(a) and 15A-1340.17(c)(2)]. A judge, however, can deviate upward from the presumptive range and sentence in the aggravated range if he or she determines that aggravating factors are sufficient to outweigh any mitigating factors [G.S. 15A-1340.16(b)]. Aggravating factors include things such as the fact that the crime was especially heinous, atrocious, or cruel [G.S. 15A-1340.16(d)(7)]. Mitigating factors include things such as the fact that the defendant has a positive employment history or is gainfully employed [G.S. 15A-1340.16(e)(19)]. A judge can sentence below the presumptive range in the mitigated range if he or she determines that mitigating factors are present and are sufficient to outweigh any aggravating factors [G.S. 15A-1340.16(b)].

If the court selects a term from the aggravated or mitigated range, the court must make written findings of the aggravating and mitigating factors. The court must make such findings regardless of whether it imposes an active or suspended term of imprisonment [G.S. 15A-1340.16(c)].

Before the *Blakely* decision, discussed above, the judge not only weighed the aggravating and mitigating factors but also determined whether any aggravating or mitigating factors were present. After *Blakely*, aggravating factors must be submitted to the jury and proved beyond a reasonable doubt, unless properly admitted by the defendant (359 N.C. 425). Although S.L. 2005-145 carved out an exception for aggravating factor G.S. 15A-1340.16(d)(18a) (involving certain prior delinquency adjudications), the North Carolina Court of Appeals subsequently held that this factor must be submitted to the jury and proved beyond a reasonable doubt (172 N.C. App. 135). *Blakely* did not affect the sentencing judge's authority to find mitigating factors or weigh aggravating and mitigating factors (359 N.C. 425). For more detail on these issues, see Jessica Smith, *North Carolina Sentencing After* Blakely v. Washington *and the* Blakely *Bill* (September 2005), available at www.sog.unc.edu/programs/crimlaw/Blakely%20Update.pdf.

Minimum Sentence

Once the class of felony, prior record level, and aggravating and mitigating factors are determined, the court must select a minimum term of imprisonment from Table 1. To determine the minimum term, the court must locate the class of felony at issue along the left-hand side of the grid and the prior record level along the top of the grid. The cell in which the felony class and prior record level intersect shows the possible sentences that the court may impose. The court then must determine whether to impose a term of imprisonment from the presumptive, aggravated, or mitigated ranges (expressed in months) shown in the particular cell. Lastly, the court must select a minimum term of imprisonment from within the applicable range.

Maximum Sentence

The judgment of the court also must contain a maximum term of imprisonment [G.S. 15A-1340.13(c)]. The maximum term is set by statute based on the minimum term imposed by the court. For Class B1 through E felonies, the maximum term of imprisonment is 120 percent of the minimum term rounded to the next highest month, plus nine months. For Class F through I felonies, the maximum term of imprisonment is 120 percent of the minimum term rounded to the next highest month [G.S. 15A-1340.17(d), (e), (e1)].

A minimum/maximum table containing these calculations is shown as Table 2. The numbers in the table to the left of the dash represent the minimum term of imprisonment imposed by the court, expressed in months. The numbers to the right of the dash represent the corresponding maximum term required by statute. The table is in two parts—the first part lists maximum sentences for Class B1 through E felonies, and the second lists the maximum sentences for Class F through I felonies.

Sentence Disposition

The next step is to determine the sentence disposition, which is prescribed in Table 1. Each cell in the grid contains a sentence disposition, signified by the letter "A," "I," or "C," or a combination of these letters. "A" represents active punishment (jail or prison); "I" represents intermediate punishment (a form of probation—more strict supervision than unsupervised probation but less restrictive than jail or prison); and "C" represents community punishment (the least restrictive form of probation) [G.S. 15A-1340.17(c)(1)]. The court must impose the sentence disposition indicated in the applicable cell. Some cells in the grid prescribe two possible dispositions, separated by a slash (for example, "I/A"). In those cases, the court can impose either disposition.

Active punishment. This is an unsuspended term of imprisonment (G.S. 15A-1340.11). If the court imposes an active punishment, the minimum and maximum term of imprisonment previously determined by the court may not be suspended. The maximum term may be reduced by earned-time credit (awarded by the Department of Correction or local jail), but the term of imprisonment may not be reduced (except for credit for time served awaiting trial) below the minimum term imposed by the court [G.S. 15A-1340.13(d)].[1]

---

1. A defendant convicted of a Class B1 through E felony is automatically released from prison nine months before the end of his or her maximum term of imprisonment and is placed on post-release supervision, except for B1 felons sentenced to life imprisonment without parole. The defendant may be released sooner if he or she receives any earned-time credit but may not be released before serving the minimum term of

Ordinarily, if the only disposition prescribed in a particular cell is "A," the court must impose active imprisonment. Upon a finding of extraordinary mitigation, however, the court may impose an intermediate punishment even when only an active punishment is prescribed [G.S. 15A-1340.13(g)]. Extraordinary mitigation is authorized only when the offense is a Class B2, C, or D felony; the offense is not a drug-trafficking offense under G.S. 90-95(h) or a drug-trafficking conspiracy offense under G.S. 90-95(i); *and* the defendant is in prior record level I or II [G.S. 15A-1340.13(h)].

In cases involving drug-trafficking offenses, the court is not required to impose an active punishment (regardless of the class of offense or the defendant's prior record level) if the court finds that the defendant provided "substantial assistance" within the meaning of G.S. 90-95(h)(5). Drug trafficking is discussed further below, under the heading "Special Provisions."

When sentencing a defendant for multiple offenses, the court may consolidate sentences, run them concurrently, or run them consecutively (G.S. 15A-1340.15 and 15A-1354). Unless otherwise specified by the court, sentences run concurrently [G.S. 15A-1340.15(a) and 15A-1354(a)]. If the court consolidates offenses for sentencing, the most serious offense is controlling—the sentence disposition and the minimum and maximum terms of imprisonment must conform to the structured sentencing rules for that offense [G.S. 15A-1340.15(b)]. If the court imposes consecutive sentences, the minimum term of imprisonment is the sum of the minimum terms imposed for the offenses, and the maximum term is the sum of the maximum terms for the offenses [G.S. 15A-1354(b)].[2]

A defendant sentenced to active punishment for a felony is normally committed to the custody of the Department of Correction. However, upon request of the sheriff or board of county commissioners, the court may, in its discretion, sentence the person to a local confinement facility in that county [G.S. 15A-1352(b)].

---

imprisonment imposed by the court (absent credit for time served) (G.S. 15A-1368.1; G.S. 15A-1368.2). The period of post-release supervision is nine months for most people convicted of a Class B1 through E felony. But a defendant is subject to a far longer period of supervised release—five years—if he or she has been convicted of a Class B1 through E felony and is required to register as a sex offender under G.S. Chapter 14, Article 27A [G.S. 15A-1368.2(c)]. If the defendant violates a condition of post-release supervision during that period, he or she can be returned to prison "up to the time remaining on his [or her] maximum imposed term" [G.S. 15A-1368.3(c)(1)].

2. If the court imposes consecutive terms of imprisonment for more than one Class B1 through E felony, the maximum term for each second and subsequent Class B1 through E felony is reduced by nine months [G.S. 15A-1354(b)(1)].

Intermediate punishment. This is supervised probation involving at least one of the following: (1) special probation, (2) assignment to a residential program, (3) house arrest with electronic monitoring, (4) intensive probation, (5) assignment to a day-reporting center, or (6) assignment to a drug treatment court program [G.S. 15A-1340.11(6)].

If the court imposes an intermediate punishment, it must suspend the minimum and maximum term of imprisonment and impose a period of supervised probation with at least one of the conditions described in G.S. 15A-1340.11(6). For intermediate punishments for felonies, the court is authorized to impose a period of probation ranging from 18 to 36 months; the court may depart from this range upon finding that a longer or shorter period is necessary [G.S. 15A-1343.2(d)].

As a condition of intermediate punishment, the court is authorized to impose special probation, also known as a split sentence. Under special probation, the court suspends the term of imprisonment, places the defendant on probation, and requires the defendant to submit to a period of imprisonment as a condition of probation [G.S. 15A-1351(a)]. The period of imprisonment pursuant to special probation may not exceed one-fourth of the maximum term of imprisonment imposed [G.S. 15A-1351(a)].

Community punishment. Community punishment is any sentence that does not include an active or intermediate punishment [G.S. 15A-1340.11(2)]. The court must suspend any term of imprisonment; it may not impose an active term of imprisonment or special probation requiring a period of imprisonment.

A community punishment may include unsupervised probation or supervised probation with any authorized condition other than one defined as an intermediate punishment [G.S. 15A-1340.11(2) and 15A-1340.11(6)]. For community punishments for felonies, the court is authorized to impose a period of probation ranging from 12 to 30 months; the court may depart from this range upon finding that a longer or shorter period is necessary [G.S. 15A-1343.2(d)(3)]. A community punishment also may consist of a fine only, without probation [G.S. 15A-1340.17(b)].

G.S. 15A-1382.1(b) provides that if the court finds that there was a personal relationship [as defined in G.S. 50B-1(b)] between the defendant and the victim and imposes a sentence of community punishment, the court must determine whether the defendant must comply with the special conditions of probation in G.S. 15A-1343(b1). That subsection also provides that the court may impose house arrest under G.S. 15A-1343(b1)(3c), even though such a condition is authorized in other cases only if the court imposes intermediate punishment. It is not clear whether this subsection is meant to apply to any case in which the court finds that there was a personal relationship or whether it is limited to the offenses mentioned in G.S. 15A-1382.1(a), which include assault or communicating a threat.

Fines. The court may impose a fine as part of any disposition, whether active, intermediate, or community. Unless otherwise provided by statute, the amount of the fine is in the court's discretion [G.S. 15A-1340.17(b), 15A-1340.23(b), and 15A-1343(b)(9)].

Restitution

The last step in felony sentencing is to consider the appropriateness of restitution. Article 81C of G.S. Chapter 15A (G.S. 15A-1340.34 through 15A-1340.38) governs restitution in all criminal cases. The restitution requirements differ, however, depending on whether or not the offense is subject to the Crime Victims' Rights Act (G.S. 15A-830 through 15A-841). The felonies subject to the Crime Victims' Rights Act are (1) any Class A through E felony, (2) a Class F through I felony if it is in violation of certain statutes, or (3) an attempt to commit one of the above felonies if the attempt is punishable as a felony.

The discussion below outlines the main differences concerning restitution between offenses subject to the Crime Victims' Rights Act and other criminal offenses. Note that a statute governing a particular offense may contain more specific restitution requirements.

For offenses subject to the Crime Victims' Rights Act, the court must order restitution to the victim or victim's estate [G.S. 15A-1340.34(b)]. If the defendant is placed on probation or post-release supervision, any restitution ordered must be included as a condition of probation or post-release supervision [G.S. 15A-1340.34(b)]. Even if the defendant is sentenced to active imprisonment, it appears that the court must order restitution to the victim or victim's estate. If a restitution order to a victim is for more than $250, it is enforceable as a civil judgment and, in some circumstances, may be subject to immediate execution (G.S. 15A-1340.38).

In cases not subject to the Crime Victims' Rights Act, the court must consider whether restitution is appropriate, but the court is not required to order it [G.S. 15A-1340.34(a), (c)]. Thus the court may make restitution a condition of probation. It appears that the court also may impose restitution as part of a sentence of active imprisonment. In cases not subject to the Crime Victims' Rights Act, a restitution order is not enforceable as a civil judgment.

In all cases, the victim or victim's estate may bring a civil suit for damages resulting from the crime [G.S. 15A-1340.37(a)]. Also, if the defendant is sentenced to active imprisonment, the court must consider whether to recommend to the Department of Correction that restitution be made from any work-release earnings [G.S. 15A-1340.36(c)].

The losses and injuries for which restitution may be ordered are described in G.S. 15A-1340.35. In determining the amount of restitution, the court must have adequate proof of the injuries or losses claimed and must take into account the defendant's ability to pay (G.S. 15A-1340.36). The court also may (but is not required to) order restitution to people other than the victim or to organizations (G.S. 15A-1340.37).

Special Provisions

A number of provisions depart from the basic structured sentencing scheme for felonies, described above. These departures are noted in the punishment charts where applicable. The principal exceptions are as follows:

Class A felonies. Class A felonies are punishable by death or life without parole, regardless of the defendant's prior record level. The only Class A felony is first-degree murder under G.S. 14-17. For a discussion of capital sentencing issues, see Robert L. Farb, *North Carolina Capital Case Law Handbook* (UNC School of Government, 2d ed. 2004).

Habitual felon. A person becomes a habitual felon when he or she has been convicted of three felony offenses as set out in G.S. 14-7.1. When a defendant is convicted of a felony after having achieved the status of habitual felon, the punishment for that offense is elevated to a Class C felony (unless the offense for which he or she was convicted is a Class A, B1, or B2 felony) (G.S. 14-7.6). For example, a defendant found to be a habitual felon after being convicted of felonious breaking or entering under G.S. 14-54 is sentenced as though he or she was convicted of a Class C felony, not a Class H felony, which is the classification of that offense. [A defendant's status as a habitual felon is determined by a jury at a hearing held after a conviction, unless the defendant admits to being a habitual felon (G.S. 14-7.5).] Prior convictions used to establish habitual felon status cannot be used in determining the prior record level in sentencing for the Class C felony (G.S. 14-7.6).

Violent habitual felon. A person becomes a violent habitual felon when he or she has been convicted of two violent felony offenses as set out in G.S. 14-7.7. When a defendant is convicted of a violent felony after having achieved the status of violent habitual felon, the punishment for that offense (except when the death penalty has been imposed) is life imprisonment without parole (G.S. 14-7.12).

Firearm enhancement. Subject to certain exceptions, a defendant who, during the commission of a Class A through E felony, (1) used, displayed, or threatened to use or display a firearm and (2) actually possessed the firearm about his or her person must be sentenced to an additional 60 months imprisonment [G.S. 15A-1340.16A(c)]. The 60-month enhancement attaches to the minimum term of imprisonment; the applicable maximum term is calculated using the enhanced minimum [G.S. 15A-1340.16A(c)]. The facts supporting the enhancement must be alleged in an indictment or information and proved to a jury beyond a reasonable doubt, unless the defendant pleads guilty or no contest to the issue [G.S. 15A-1340.16A(d), (e); 353 N.C. 568].

B1 felonies against young victims. Subject to an exception, a person who commits a Class B1 felony—such as first-degree rape or sex offense—against a victim who was 13 years old or younger and has a prior Class B1 felony conviction will be sentenced to life imprisonment without parole [G.S. 15A-1340.16B(a)]. The facts supporting the enhancement must be alleged in an indictment or information and proved to a jury beyond a reasonable doubt, unless the defendant pleads guilty or no contest to the issue [G.S. 15A-1340.16B(d), (e)].

Bullet-proof vest enhancement. Subject to certain exceptions, a person who commits a felony while wearing or having in his or her immediate possession a bullet-proof vest will be punished one class higher than the underlying felony [G.S. 15A-1340.16C(a)]. Thus, if the felony is punishable as a Class D felony, the person will be sentenced as a Class C felon. The facts supporting the enhancement must be alleged in an indictment or information and proved to a jury beyond a reasonable doubt, unless the defendant pleads guilty or no contest to the issue [G.S. 15A-1340.16C(c), (d)].

Methamphetamine enhancement. Unless an exception applies, if a person is convicted of manufacturing of methamphetamine and a law enforcement, probation, or parole officer, emergency medical services employee, or firefighter suffered serious injury caused by the hazards associated with the manufacture of methamphetamine, then that person's minimum term of imprisonment is enhanced by 24 months [G.S. 15A-1340.16D(a)]. The facts supporting the enhancement must be alleged in an indictment or information and proved to a jury beyond a reasonable doubt, unless the defendant pleads guilty or no contest to the issue [G.S. 15A-1340.16D(b), (c)].

Drug trafficking. Drug trafficking is punished according to a separate table of punishments, containing minimum and maximum terms of imprisonment that depart from the sentencing grid. Minimum fines also are prescribed for drug-trafficking offenses. See G.S. 90-95(h) and trafficking offenses in Chapter 27.

Misdemeanors

The focus of structured sentencing for misdemeanors is the statutory table of punishments in G.S. 15A-1340.23 (see Table 4 at the end of this chapter). All misdemeanors, except those mentioned at the beginning of this chapter, are subject to structured sentencing (G.S. 15A-1340.10).

A sentencing court essentially must take five steps to determine the appropriate sentence for a misdemeanor. These steps (with the appropriate source to consult) are as follows:

(1) Determine the offense class for each misdemeanor conviction (listed under each offense in this book as "Punishment")
(2) Determine the prior conviction level for the defendant (Table 3)
(3) Select a sentence length from the appropriate sentence range (Table 4)
(4) Determine the sentence disposition (active, intermediate, or community) (Table 4)
(5) Consider the appropriateness of restitution (see the discussion later in this chapter)

The sentencing court will inform the defendant of (1) the active jail or prison term imposed and (2) whether that term will be activated or suspended. If the active term is suspended, the defendant will be informed of the type of alternative disposition (intermediate or community) he or she will receive. This procedure will be followed for each offense conviction, unless a particular conviction is consolidated or set to run concurrently with other convictions.

The *Blakely* decision discussed above has no impact on structured sentencing for misdemeanors. Under structured sentencing, the only enhancing factors that apply in misdemeanor sentencing are prior convictions—factors specifically excluded from *Blakely.*

Offense Class

The first step in determining the appropriate sentence for a misdemeanor is to identify the class of the misdemeanor. There are four classes of misdemeanors under structured sentencing: A1, 1, 2, and 3. Punishments increase in severity as you move from a Class 3 misdemeanor up to a Class A1 misdemeanor. For each misdemeanor listed in this book, the class of the offense is indicated under the "Punishment" section of the discussion. For example, "Assault by Pointing a Gun" (G.S. 14-34; see Chapter 7, "Assaults") is a Class A1 misdemeanor.

Some misdemeanor offenses have no classification and no punishment listed in the General Statutes. Under G.S. 14-3(a), those offenses generally are considered Class 1 misdemeanors. Misdemeanors that are "infamous, done in secrecy and malice, or with deceit and intent to defraud" are punished as Class H felonies pursuant to G.S. 14-3(b).[3] Some misdemeanor offenses have a punishment

---

3. For more detail on this issue, see "Committing an Infamous or Related Misdemeanor" in Jessica Smith, *North Carolina Crimes: A Guidebook on the Elements of Crime*, 6th ed. (Chapel Hill: UNC School of Government, 2007), 59–60.

but no classification listed. Under G.S. 14-3(a), those offenses are classified as follows: as a Class 1 misdemeanor if punishable by more than six months imprisonment; as a Class 2 misdemeanor if punishable by more than 30 days but not more than six months imprisonment; and as a Class 3 misdemeanor if punishable by imprisonment of 30 days or less or by a fine only.

Prior Conviction Level

The second step in determining the appropriate sentence for a misdemeanor is to determine the defendant's prior conviction level. A defendant is assigned to one of three prior conviction levels (I through III) based on his or her total number of prior felony and misdemeanor convictions [G.S. 15A-1340.21(b)]. The three prior conviction levels, and the number of convictions applicable to each level, are indicated in Table 4.

Any conviction, whether a felony or misdemeanor (including driving while impaired and other misdemeanors under Chapter 20 of the North Carolina General Statutes), counts as one conviction [G.S. 15A-1340.21(b)]. If the defendant was convicted of more than one offense in a single week of superior court or a single session (that is, day) of district court, only one of the convictions counts [G.S. 15A-1340.21(d)]. Infractions do not count.

A prior offense may be counted as a conviction only if the offense is classified as a felony or misdemeanor at the time the defendant committed the current offense [G.S. 15A-1340.21(b)]. Thus, if an offense had been changed from a misdemeanor to an infraction (for example, speeding 50 m.p.h. in a 35 m.p.h. zone) when a defendant committed a new misdemeanor, a prior conviction for that offense would not count in misdemeanor sentencing.

Sentence Length

Once the class of misdemeanor and prior conviction level are determined, the court must determine the length of any term of imprisonment. (If the court selects a community punishment as the sentence disposition, discussed under the next heading, it may impose a judgment consisting of a fine only; in those circumstances, it would be unnecessary for the court to specify any term of imprisonment.) To determine the length of any term of imprisonment, the court must locate the class of misdemeanor at issue along the left-hand side of Table 4 and the prior conviction level along the top of Table 4. The cell in which the misdemeanor class and prior conviction level intersect shows the possible terms of imprisonment (expressed in days) that the court may impose. The court must select a single term of imprisonment from the range shown in the applicable cell; there are no minimum and maximum terms of imprisonment, as in felony sentencing.

Sentence Disposition

The next step in misdemeanor sentencing is to determine the sentence disposition, which is prescribed in Table 4. Each cell in Table 4 contains a sentence disposition, signified by the letter "A," "I," or "C," or a combination of these letters. "A" represents active punishment (jail or prison); "I" represents intermediate punishment (a form of probation—more strict supervision than unsupervised probation but less restrictive than jail or prison); and "C" represents community punishment (the least restrictive form of probation) [G.S. 15A-1340.23(c)]. The court must impose the sentence disposition indicated in the applicable cell. Some cells prescribe more than one possible disposition, separated by a slash (for example, "C/I/A"). In those cases, the court can impose any single one of the indicated dispositions.

Active punishment. If the court imposes an active punishment, the term of imprisonment previously determined by the court must be activated. A defendant's term of imprisonment may be reduced by earned-time credit up to four days per month of incarceration (awarded by the Department of Correction or local jail) [G.S. 15A-1340.20(d)].

When sentencing a defendant for multiple offenses, the court may consolidate sentences or run them concurrently. Subject to certain limitations, the court also may impose consecutive sentences. Unless otherwise specified by the court, sentences run concurrently. If the court consolidates offenses for sentencing, the most serious offense is controlling—the sentence disposition and the term of imprisonment must conform to the structured sentencing rules for that offense [G.S. 15A-1340.22(b)]. If the court imposes consecutive sentences, the length of imprisonment cannot exceed twice the longest term of imprisonment authorized for the most serious misdemeanor conviction [G.S. 15A-1340.22(a)]. Consecutive sentences cannot be imposed, however, if all of the convictions are for Class 3 misdemeanors [G.S. 15A-1340.22(a)].

A defendant sentenced for a misdemeanor to active punishment of 90 days or less must be committed to a local jail facility [G.S. 15A-1352(a)]. The court may choose local confinement or commitment to the custody of the Department of Correction if the defendant is sentenced to more than 90 days of active punishment [G.S. 15A-1352(a)].

An active sentence may be imposed for any misdemeanor, even if an active sentence would not otherwise be authorized, if the sentence does not exceed the total amount of time that the defendant has spent in pretrial confinement awaiting trial for that misdemeanor. In effect, a sentence of "credit for time served" may be imposed for any misdemeanor [G.S. 15A-1340.20(c1)].

Intermediate punishment. Intermediate punishment for misdemeanors is the same as intermediate punishment for felonies, except for the period of probation. For misdemeanor intermediate punishment, the court is authorized to impose a period of probation ranging from 12 to 24 months; the court may depart from this range upon finding that a longer or shorter period is necessary [G.S. 15A-1343.2(d)(2)].

Community punishment. Community punishment for misdemeanors is the same as community punishment for felonies, except for the period of probation. For misdemeanor community punishment, the court is authorized to impose a period of probation ranging from 6 to 18 months; the court may depart from this range upon finding that a longer or shorter period is necessary [G.S. 15A-1343.2(d)(1)]. A community punishment also may consist of a fine only, without probation [G.S. 15A-1340.23(b)].

Fines. The court may impose a fine as part of any disposition, whether active, intermediate, or community [G.S. 15A-1340.23(b)]. Unless otherwise provided by statute, the maximum fine for each class of misdemeanor is as indicated in Table 4.

Restitution

The last step in misdemeanor sentencing is to determine the appropriateness of restitution. The restitution requirements in G.S. 15A-1340.34 through 15A-1340.38 (discussed earlier in this chapter in connection with felonies) apply equally to misdemeanors.

The main difference with restitution in misdemeanor versus felony cases is that far fewer misdemeanors are subject to the Crime Victims' Rights Act. Only the following misdemeanors are covered: (1) assault with a deadly weapon, (2) assault inflicting serious injury, (3) assault on a female, (4) simple assault, (5) assault by pointing a gun, (6) domestic criminal trespass, and (7) stalking [G.S. 15A-830(7)(g)]. Further, the Crime Victims' Rights Act applies to the above misdemeanors only if the defendant and victim were in one of six different "personal relationships" (for example, as current or former spouses) described in G.S. 50B-1(b) [G.S. 15A-830(7)(g)].

Special Provisions

Infamous or related misdemeanor enhancement. G.S. 14-3(b) provides that if a person commits a misdemeanor for which no specific punishment is prescribed and the misdemeanor is infamous, done in secrecy and malice, or done with deceit and intent to defraud, punishment is elevated to a Class H felony.[4]

Prejudice enhancement. G.S. 14-3(c) provides for an enhanced punishment if a misdemeanor is committed because of the victim's race, color, religion, nationality, or country of origin.[5]

---

4. See note 3 above.
5. For more detail, see "Committing a Misdemeanor Because of Prejudice," *id.*, 60–61.

Table 1
Felony Sentence Dispositions and Minimum Prison/Jail Term Ranges (shown in months)

A = active punishment **I** = intermediate punishment **C** = community punishment

Felony Class	I 0 Pts	II 1–4 Pts	III 5–8 Pts	IV 9–14 Pts	V 15–18 Pts	VI 19+ Pts	
A			**Death or life without parole**				
B1	**A**	**A**	**A**	**A**	**A**	**A**	**Disposition**
	240–300	288–360	336–420	384–480	Life without parole	Life without parole	*Aggravated Range*
	192–240	230–288	269–336	307–384	346–433	384–480	Presumptive Range
	144–192	173–230	202–269	230–307	260–346	288–384	*Mitigated Range*
B2	**A**	**A**	**A**	**A**	**A**	**A**	
	157–196	189–237	220–276	251–313	282–353	313–392	*Aggravated Range*
	125–157	151–189	176–220	201–251	225–282	251–313	Presumptive Range
	94–125	114–151	132–176	151–201	169–225	188–251	*Mitigated Range*
C	**A**	**A**	**A**	**A**	**A**	**A**	
	73–92	100–125	116–145	133–167	151–188	168–210	*Aggravated Range*
	58–73	80–100	93–116	107–133	121–151	135–168	Presumptive Range
	44–58	60–80	70–93	80–107	90–121	101–135	*Mitigated Range*
D	**A**	**A**	**A**	**A**	**A**	**A**	
	64–80	77–95	103–129	117–146	133–167	146–183	*Aggravated Range*
	51–64	61–77	82–103	94–117	107–133	117–146	Presumptive Range
	38–51	46–61	61–82	71–94	80–107	88–117	*Mitigated Range*
E	**I/A**	**I/A**	**A**	**A**	**A**	**A**	
	25–31	29–36	34–42	46–58	53–66	59–74	*Aggravated Range*
	20–25	23–29	27–34	37–46	42–53	47–59	Presumptive Range
	15–20	17–23	20–27	28–37	32–42	35–47	*Mitigated Range*
F	**I/A**	**I/A**	**I/A**	**A**	**A**	**A**	
	16–20	19–24	21–26	25–31	34–42	39–49	*Aggravated Range*
	13–16	15–19	17–21	20–25	27–34	31–39	Presumptive Range
	10–13	11–15	13–17	15–20	20–27	23–31	*Mitigated Range*
G	**I/A**	**I/A**	**I/A**	**I/A**	**A**	**A**	
	13–16	15–19	16–20	20–25	21–26	29–36	*Aggravated Range*
	10–13	12–15	13–16	16–20	17–21	23–29	Presumptive Range
	8–10	9–12	10–13	12–16	13–17	17–23	*Mitigated Range*
H	**C/I/A**	**I/A**	**I/A**	**I/A**	**I/A**	**A**	
	6–8	8–10	10–12	11–14	15–19	20–25	*Aggravated Range*
	5–6	6–8	8–10	9–11	12–15	16–20	Presumptive Range
	4–5	4–6	6–8	7–9	9–12	12–16	*Mitigated Range*
I	**C**	**C/I**	**I**	**I/A**	**I/A**	**I/A**	
	6–8	6–8	6–8	8–10	9–11	10–12	*Aggravated Range*
	4–6	4–6	5–6	6–8	7–9	8–10	Presumptive Range
	3–4	3–4	4–5	4–6	5–7	6–8	*Mitigated Range*

Source: G.S. 15A-1340.17(c). Class B2, C, and D ranges are for offenses committed on or after December 1, 1995. For shorter ranges applying before that date, see the repealed version of this statute.

Table 2
Felony Minimum and Maximum Prison/Jail Terms (shown in months)

Felony Classes B1, B2, C, D, and E

15–27	56–77	97–126	138–175	179–224	220–273	261–323	302–372
16–29	57–78	98–127	139–176	180–225	221–275	262–324	303–373
17–30	58–79	99–128	140–177	181–227	222–276	263–325	304–374
18–31	59–80	100–129	141–179	182–228	223–277	264–326	305–375
19–32	60–81	101–131	142–180	183–229	224–278	265–327	306–377
20–33	61–83	102–132	143–181	184–230	225–279	266–329	307–378
21–35	62–84	103–133	144–182	185–231	226–281	267–330	308–379
22–36	63–85	104–134	145–183	186–233	227–282	268–331	309–380
23–37	64–86	105–135	146–185	187–234	228–283	269–332	310–381
24–38	65–87	106–137	147–186	188–235	229–284	270–333	311–383
25–39	66–89	107–138	148–187	189–236	230–285	271–335	312–384
26–41	67–90	108–139	149–188	190–237	231–287	272–336	313–385
27–42	68–91	109–140	150–189	191–239	232–288	273–337	314–386
28–43	69–92	110–141	151–191	192–240	233–289	274–338	315–387
29–44	70–93	111–143	152–192	193–241	234–290	275–339	316–389
30–45	71–95	112–144	153–193	194–242	235–291	276–341	317–390
31–47	72–96	113–145	154–194	195–243	236–293	277–342	318–391
32–48	73–97	114–146	155–195	196–245	237–294	278–343	319–392
33–49	74–98	115–147	156–197	197–246	238–295	279–344	320–393
34–50	75–99	116–149	157–198	198–247	239–296	280–345	321–395
35–51	76–101	117–150	158–199	199–248	240–297	281–347	322–396
36–53	77–102	118–151	159–200	200–249	241–299	282–348	323–397
37–54	78–103	119–152	160–201	201–251	242–300	283–349	324–398
38–55	79–104	120–153	161–203	202–252	243–301	284–350	325–399
39–56	80–105	121–155	162–204	203–253	244–302	285–351	326–401
40–57	81–107	122–156	163–205	204–254	245–303	286–353	327–402
41–59	82–108	123–157	164–206	205–255	246–305	287–354	328–403
42–60	83–109	124–158	165–207	206–257	247–306	288–355	329–404
43–61	84–110	125–159	166–209	207–258	248–307	289–356	330–405
44–62	85–111	126–161	167–210	208–259	249–308	290–357	331–407
45–63	86–113	127–162	168–211	209–260	250–309	291–359	332–408
46–65	87–114	128–163	169–212	210–261	251–311	292–360	333–409
47–66	88–115	129–164	170–213	211–263	252–312	293–361	334–410
48–67	89–116	130–165	171–215	212–264	253–313	294–362	335–411
49–68	90–117	131–167	172–216	213–265	254–314	295–363	336–413
50–69	91–119	132–168	173–217	214–266	255–315	296–365	337–414
51–71	92–120	133–169	174–218	215–267	256–317	297–366	338–415
52–72	93–121	134–170	175–219	216–269	257–318	298–367	339–416
53–73	94–122	135–171	176–221	217–270	258–319	299–368	340 or more*
54–74	95–123	136–173	177–222	218–271	259–320	300–369	
55–75	96–125	137–174	178–223	219–272	260–321	301–371	

Table 2 (*continued*)
Felony Minimum and Maximum Prison/Jail Terms
(shown in months)

Felony Classes F, G, H, and I

3–4	9–11	15–18	21–26	27–33	33–40	39–47	45–54
4–5	10–12	16–20	22–27	28–34	34–41	40–48	46–56
5–6	11–14	17–21	23–28	29–35	35–42	41–50	47–57
6–8	12–15	18–22	24–29	30–36	36–44	42–51	48–58
7–9	13–16	19–23	25–30	31–38	37–45	43–52	49–59
8–10	14–17	20–24	26–32	32–39	38–46	44–53	

* Where minimum term is 340 months or more, maximum is 120% of minimum rounded to next highest month, plus 9 months.

Source: G.S. 15A-1340.17(e), (e1).

Table 3

STATE OF NORTH CAROLINA

File No.

____________________ County

In The General Court Of Justice
☐ District ☐ Superior Court Division

STATE VERSUS

Name And Address Of Defendant

Social Security No | SID No.

Race | Sex | DOB

WORKSHEET
PRIOR RECORD LEVEL FOR FELONY SENTENCING AND PRIOR CONVICTION LEVEL FOR MISDEMEANOR SENTENCING
(STRUCTURED SENTENCING)

G.S. 15A-1340.14, 15A-1340.21

I. SCORING PRIOR RECORD/FELONY SENTENCING

NUMBER	TYPE	FACTORS	POINTS
	Prior Felony Class A Conviction	X 10	
	Prior Felony Class B1 Conviction	X 9	
	Prior Felony Class B2 or C or D Conviction	X 6	
	Prior Felony Class E or F or G Conviction	X 4	
	Prior Felony Class H or I Conviction	X 2	
	Prior Class A1 or 1 Misdemeanor Conviction *(see note on reverse)*	X 1	
		SUBTOTAL ▶	

Defendant's Current Charge(s):

If all the elements of the present offense are included in any prior offense whether or not the prior offenses were used in determining prior record level.	+ 1	
If the offense was committed: (a) while on supervised or unsupervised probation, parole, or post-release supervision; or (b) while serving a sentence of imprisonment; or (c) while on escape.	+ 1	
	TOTAL ▶	

II. CLASSIFYING PRIOR RECORD/CONVICTION LEVEL

MISDEMEANOR

NOTE: *If sentencing for a misdemeanor, total the number of prior conviction(s) listed on the reverse and select the corresponding prior conviction level.*

No. Of Prior Convictions	Level
0	I
1 - 4	II
5+	III

PRIOR CONVICTION LEVEL ▶ ☐

☐ The Court has determined the number of prior convictions to be __________ and the level to be as shown above.

☐ In making this determination, the Court has relied upon the State's evidence of the defendant's prior convictions from a computer printout of DCI-CCH.

FELONY

NOTE: *If sentencing for a felony, locate the prior record level which corresponds to the total points determined in Section I above.*

Points	Level
0	I
1 - 4	II
5 - 8	III
9 - 14	IV
15 - 18	V
19+	VI

PRIOR RECORD LEVEL ▶ ☐

☐ The Court finds the prior convictions, prior record points and the prior record level of the defendant to be as shown herein.

☐ In making this determination, the Court has relied upon the State's evidence of the defendant's prior convictions from a computer printout of DCI-CCH.

☐ In finding a prior record level point under G.S. 15A-1340.14(b)(7), the Court has relied on the jury's determination of this issue beyond a reasonable doubt or the defendant's admission to this issue.

☐ For each out-of-state conviction listed in Section IV on the reverse, the Court finds by a preponderance of the evidence that the offense is substantially similar to a North Carolina offense and that the North Carolina classification assigned to this offense in Section IV is correct.

☐ The Court finds that the State and the defendant have stipulated in open court to the prior convictions, points and record level.

Date	Name Of Presiding Judge (Type Or Print)	Signature Of Presiding Judge

AOC-CR-600, Rev. 10/06 (Over)

III. STIPULATION

The prosecutor and defense counsel, or the defendant, if not represented by counsel, stipulate to the information set out in Sections I and IV of this form, and agree with the defendant's prior record level or prior conviction level as set out in Section II based on the information herein.

Date	Signature Of Prosecutor	Date	Signature Of Defense Counsel Or Defendant

IV. PRIOR CONVICTION

NOTE: *Federal law precludes making computer printout of DCI-CCH (rap sheet) part of permanent public court record.*

NOTE: *The only Class 1 misdemeanor offenses under Chapter 20 that are assigned points for determining prior record level for felony sentencing are misdemeanor death by vehicle [G.S. 20-141.4(a2)] and, for sentencing for felony offenses committed on or after December 1, 1997, impaired driving [G.S. 20-138.1] and commercial impaired driving [G.S. 20-138.2]. First Degree Rape and First Degree Sexual Offense convictions prior to October 1, 1994, are Class B1 convictions.*

Source Code	Offenses	File No.	Date Of Conviction	County (*Name of State if not NC*)	Class

Source Code: 1 - DCI; 2 - NCIC; 3 - AOC/Local; 4 - AOC/Statewide; 5 - ID Bureau; 6 - Other

Date Prepared: ____________________

Prepared By: ____________________

AOC-CR-600, Side Two, Rev. 10/06

Table 4
Misdemeanor Sentence Dispositions and Prison/Jail Term Ranges

A = active punishment **I** = intermediate punishment **C** = community punishment

	Level I: No Prior Convictions	Level II: 1 to 4 Prior Convictions	Level III: 5 or More Prior Convictions
Misdemeanor Class A1 (fine discretionary)	C/I/A 1–60 days	C/I/A 1–75 days	C/I/A 1–150 days
Misdemeanor Class 1 (fine discretionary)	C 1–45 days	C/I/A 1–45 days	C/I/A 1–120 days
Misdemeanor Class 2 (maximum fine $1,000)	C 1–30 days	C/I 1–45 days	C/I/A 1–60 days
Misdemeanor Class 3 (maximum fine $200)	C 1–10 days	C/I 1–15 days	C/I/A 1–20 days

Source: G.S. 15A-1340.23(b), (c).

ml 9/11